'This book is an important contribution in what is a key area of safety science within health care organizations – the field of resilient health care organizations. It presents interesting and novel ideas and also a powerful way of thinking about good practice which should be timely and relevant for front line clinicians, nurses and health managers alike.'

– *Ewan Ferlie*, Professor of Public Services Management,
King's College London, UK

'Medicine has been late to shift from a deterministic view of the universe to an emergent and adaptive one. This fourth book in a must-read series adds depth, detail and worked examples which illustrate why this shift is both important and urgent.'

– *Trish Greenhalgh*, Professor of Primary Care Health Sciences,
University of Oxford, UK

'We need to deliver value from health care for ageing populations, with multi-morbidity, increasingly complex health care capability in which less may be more, emphasis on patient experience, and finite budgets in which we need to do more with less. These resilient health care strategies show us new ways to tackle these challenges.'

– *Adrian Edwards*, Professor of General Practice,
Cardiff University, UK

'As ideas from resilience engineering re-shape our thinking about health care, it is timely that new methodological insights also inform how researchers and practitioners apply these ideas to their improvement work. Hollnagel, Braithwaite and Wear again bring us essential learning on the practical application of resilience engineering to health care.'

– *Justin Waring*, Professor of Organisational Sociology,
Nottingham University, UK

'The change of perspective from Safety-I to Safety-II is a major breakthrough in thinking about patient safety. Putting the relation between "work-as-imagined" and "work-as-done" center stage, this book offers tools that help both scholars and practitioners in researching and designing safer health care systems.'

– *Roland Bal*, Professor of Healthcare Governance, Erasmus
University Rotterdam, The Netherlands

Delivering Resilient Health Care

Health care is under tremendous pressure regarding efficiency, safety, and economic viability. It has responded by adopting techniques that have been useful in other industries, such as quality management, lean production, and high reliability – although with limited, and all-too-often disappointing, results. The Resilient Health Care Network (RHCN) has worked since 2011 to facilitate the interaction and collaboration among practitioners and researchers interested in applying concepts from resilience engineering to health care and patient safety. This has met with considerable success, not least because the focus from the start was on developing concrete ways to complement a Safety-I perspective with a Safety-II perspective.

Building on previous volumes, *Delivering Resilient Health Care* presents documented experiences and practical guidance on how to bring Resilient Health Care into practice. It provides concrete advice on how to prepare a study, how to choose the right data, how to collect it, how to analyse the data, and how to interpret the results. This fourth book in the Resilient Health Care series contains contributions from international experts in health care, organisational studies and patient safety, as well as resilience engineering.

This book provides a practical guide for delivering resilient health care, particularly for clinicians on the frontline of care unsure how to incorporate resilience into their everyday work, managers coordinating care, and for policymakers hoping to steer the system in the right direction. Other groups – patients, the media, and researchers – will also find much of interest here.

Erik Hollnagel is Senior Professor of Patient Safety at Jönköping University (Sweden). He has worked at universities, research centres, and with industries in many countries and with such wide-ranging issues as nuclear power generation, aerospace and aviation, software engineering, land-based and maritime transportation, industrial production, and health care.

Jeffrey Braithwaite, Professor of Health Systems Research at Macquarie University (Australia), Founding Director of the Australian Institute of Health Innovation, and Director of the Centre for Healthcare Resilience and Implementation Science, is a leading health services and systems researcher with an international reputation for his work.

Robert L. Wears, Professor in Emergency Medicine at the University of Florida (USA), was a senior physician and leading international expert in patient safety. He was a visiting Professor at the Imperial College London (UK), senior associate dean for hospital affairs, and former chair of emergency medicine.

Delivering Resilient Health Care

Edited by Erik Hollnagel,
Jeffrey Braithwaite
and Robert L. Wears

Routledge
Taylor & Francis Group

LONDON AND NEW YORK

First published 2019
by Routledge
2 Park Square, Milton Park, Abingdon, Oxon OX14 4RN

and by Routledge
711 Third Avenue, New York, NY 10017

Routledge is an imprint of the Taylor & Francis Group, an informa business

British Library Cataloguing in Publication Data
A catalogue record for this book is available from the British
Library

Library of Congress Cataloging-in-Publication Data
Names: Hollnagel, Erik, 1941- editor. | Braithwaite, Jeffrey,
1954- editor. | Wears, Robert L., editor.
Title: Delivering resilient health care / edited by Erik Hollnagel,
Jeffrey Braithwaite and Robert L. Wears.
Description: Abingdon, Oxon ; New York, NY : Routledge, 2019.
| Includes bibliographical references and index.
Identifiers: LCCN 2018021570| ISBN 9781138602243 (hardback)
| ISBN 9781138602250 (pbk.) | ISBN 9780429469695 (ebook)
Subjects: | MESH: Delivery of Health Care | Health Services
Administration | Quality Control
Classification: LCC RA418 | NLM W 84.1 | DDC 362.1--dc23
LC record available at https://lccn.loc.gov/2018021570

ISBN: 978-1-138-60224-3 (hbk)
ISBN: 978-1-138-60225-0 (pbk)
ISBN: 978-0-429-46969-5 (ebk)

Typeset in Garamond
by Integra Software Services Pvt. Ltd.

Contents

Chapter 1

Coming of age

Jeffrey Braithwaite and Erik Hollnagel

In this, the fourth volume focusing on understanding health care from a resilience engineering perspective, the focus has changed from theorising about, describing and analysing resilient health care (RHC) to examining how best to study it. Before we can articulate this, however, it might be considered wise to say a few words on how we got to here.

The first book on RHC, brought together by the community of interested parties in the Resilient Health Care Network (RHCN: www.resilienthealth care.net), offered a series of arguments and presented a number of case studies that provided the first comprehensive description of RHC (Hollnagel, Braithwaite and Wears, 2013c). After lamenting once again the difficulties that health systems face in providing care to patients and pointing out how conventional solutions were not working, the first book introduced to a health and medical audience the insight that failures and successes both 'have their origin in performance variability on individual and systemic levels' (Hollnagel, Braithwaite and Wears, 2013c, p. xxiv) and that failure is the flip side of success. The proper focus for those seeking to understand how care is and can be delivered should be the continued functioning of systems under challenging circumstances, rather than the search for and rooting out of errors and mistakes. This led us to document ideas about Safety-I and Safety-II and the complementary nature of these views and to note that Safety-I is reactive and defined as the relative absence of adverse events, whereas Safety-II is prospective and defined as the ability to succeed under varying conditions. A useful way of summarising the first book and the core concepts we were playing with across its pages is the Word Art (https://wordart.com/) or visual snapshot of the common constructs of the volume shown in Figure 1.1. The frequently recurring words highlight the essential content of the book: *resilience, care, health, patient, safety, organisation, processes, change, success, work, systems* and *performance*.

The second volume in the series (Wears, Hollnagel and Braithwaite, 2015a) homed in on the work that takes place on the frontlines of care, where patients are kept safe and programmes of activity are carried out. The dominant theme we had coalesced around in the second of the, by

Figure 1.1 Common themes from 'Resilient Health Care'.

now annual, RHCN meetings was that we needed to understand how care took place at the 'clinical coalface'. It had become abundantly clear that everyday performance is characterised by how people constantly modify what they do in order to accomplish their work. This everyday clinical work (ECW) is where things happen frequently, and it is the unfurling of daily activities of frontline clinicians that explains people's contributions to resilient health care. This is centrally about how people actually get their work done, regularly and routinely, and how they in most cases manage to keep patients safe despite all sorts of pressures and resource constraints. The Word Art in Figure 1.2 provides the key concepts from the second book. Again we see *resilience, care, health, safety* and *patient* as recurring words, but we also see aspects of the everyday on the frontlines of care, such as *practice, case, adaptive, medical, physician, nurse* and *discharge*.

By the time we got to the third volume in the series (Braithwaite, Wears and Hollnagel, 2017), we were ready to look at resilient practices from two contrasting standpoints: Work-as-Done (WAD) and Work-as-Imagined (WAI). Even a cursory inspection of a health system leads to a conclusion that WAD differs from WAI. Another way of saying this is that there will always be a gap in understanding between those who plan, prescribe, fund or mandate initiatives to keep things safe and those who treat, care for or intervene directly to alleviate patients' conditions. This is, and indeed must be, the case both logically and in practice. In any system as complex as health care, with its intricate and elaborate mix of resources, staffing categories, resource allocations, politics and professional interests, structured into hierarchies and heterarchies, there will be an inevitable separation between those

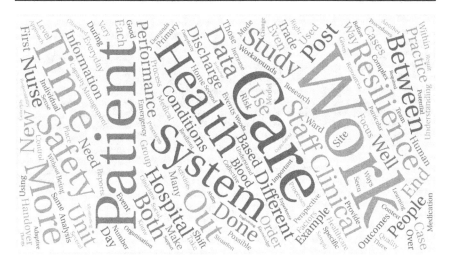

Figure 1.2 Common themes from 'The Resilience of Everyday Clinical Work'.

with global responsibility for the enterprise and those conducting operational work. The third volume explored the nature of this distinction, which in the industrial safety literature is known as the difference between what happens at the sharp end and at the blunt end. The third volume made the case that there is a need to reconcile the two world views, regardless of differences in WAI-WAD responsibilities or roles. It is essential to recognise that the issue is not whether WAD is 'right' and WAI is 'wrong', or vice versa. The reconciliation is necessary because it is impossible that clinicians doing the work of health care could carry out all the instructions, policies procedures and rules that are specified for them just as it is impossible that policy makers and managers who rely on WAI could alter the rules, policies and procedures such that they corresponded with WAD. Figure 1.3 displays the core concepts embedded in this volume, with *work, improvement, training, simulation, reporting, management*, and of course *WAD* and *WAI*, being emphasised.

Along the way to publishing this trilogy, we also released a White Paper on behalf of the RHCN such that as wide an audience as possible would have access to some of the ideas expressed in the series (Hollnagel, Wears and Braithwaite, 2015). This traced a historical argument for some of the major developments in resilience engineering (RE) from its early beginnings, and its application to health care more recently. It also contained a series of definitions for some of the constructs appearing in the RHC books, such as WAI-WAD, Safety-I and Safety-II and sharp end–blunt end distinctions. To capture the elements of this publication, we provide a fourth Word Art (Figure 1.4), with many of the terms being repeated, albeit with different emphases.

Figure 1.3 Common themes from 'Reconciling Work-as-Imagined with Work-as-Done'.

Figure 1.4 Common themes from 'From Safety-I to Safety-II: A white paper'.

Conclusion

This chapter has illustrated the short history of the Resilient Health Care Network (RHCN) by documenting at a high level the overarching interests of the corpus of RHC activities. Clearly, the RHCN has in a few years managed to set the scene, establish a body of work, and built the credentials of the network. The RHCN has, as the RHCN website (www.resilienthealthcare.net) indicates, 'facilitate[d] the interaction, and collaboration among people who are interested in applying Resilience Engineering to health care – practitioners and researchers alike'.

The next stage is to capitalise on this platform of activities, and examine the different ways that RHC can be studied for the benefit of these two groups – practitioners, and researchers. It is to this task that we now turn.

Chapter 2

The need of a guide to deliver Resilient Health Care

Erik Hollnagel and Jeffrey Braithwaite

Resilient Health Care (RHC), by the time of writing of this chapter, has reached the tender age of six. The developments so far have been summarised in Chapter 1 of this book, and can by now also be found in several other places, e.g., Braithwaite, Wears and Hollnagel (2015). While a comparison with the physiological and psychological development of a child is tempting (but misleading), it is more important to recognise that RHC during its first six years has become widely recognised as a viable supplement – and perhaps even a viable alternative – to the established approaches to safety in hospitals and clinics around the world. This mirrors the ways in which the same approach, Safety-II, has been welcomed by other industries. There are several differences between RHC and the established approaches, some major and others minor, that may explain why this has happened. The major of these are:

- The focus of RHC is on everyday clinical work and why it usually goes well (Safety-II) rather than on unpredictable adverse outcomes, such as incidents and accidents (Safety-I).
- RHC looks at work as it actually takes place (Work-as-Done) rather than at work as it is assumed or expected to be done (Work-as-Imagined). This applies to every kind of performance and for every level of the organisation – from the clinical 'coalface' to the management.
- RHC subscribes to a system-wide perspective on how safety, quality, productivity, patient satisfaction, and more, represent facets of the same reality, and on how hospitals are complex socio-technical systems rather than streamlined 'factories' for the treatment of illnesses and the 'production' of satisfied patients.

In the context of High Reliability Organisations (HRO), Weick (1987: 112) astutely noted that safety is 'invisible in the sense that reliable outcomes are constant, which means there is nothing to pay attention to'. This partly explains the preoccupation with safety in the traditional sense (Safety-I), which actually is an obsession with the lack of safety that is marked by the unsystematic but infrequent occurrence of adverse outcomes. As Reason (2000: 4) pointed out, 'safety is defined and measured more by its absence than by its presence.' When something goes wrong, when the result of an activity is significantly different from what was intended and expected, usually in the sense of being worse, it is inevitably noticed or paid attention to. But when the results are as expected, then 'nothing has happened'.

When something goes wrong there is an obvious interest in trying to understand why it happened so that we can *be* safe by 'finding and fixing' the identified causes, usually by dealing with each cause on its own. (It also helps to make us *feel* safe, by providing a socially acceptable explanation for what happened.) The focus of safety efforts and safety management in all industries, health care being no exception, is therefore on the occurrence of adverse outcomes and on finding means to ensure that their number is reduced, preferably to zero. Resilient Health Care itself started by addressing the concerns that had been expressed by the so-called patient safety movement (Wears and Sutcliffe, forthcoming). In that sense RHC has followed the same path as Resilience Engineering, which also started from the traditional concern for things that go wrong. Resilience Engineering did, however, from the very beginning emphasise that 'failures are the flip side of successes', or in other words that the outcomes we notice (as well as the outcomes that we do not notice) are produced in basically the same way regardless of whether they are acceptable or unacceptable. (This is also referred to as the principle of equivalence, cf. Hollnagel, 2012.) Work is – by definition – always done in order to bring about the intended outcomes, which in practice means that it should lead to the desired positive effects. The obvious consequence of the principle of equivalence is therefore a shift in focus from the 'unavoidable' accidents to the way that everyday activities are carried out. It is furthermore logical that an increase in the number of acceptable outcomes – the situations where work happens as it should and where outcomes therefore are acceptable – will lead to a decrease in the number of unacceptable outcomes. But this decrease importantly does not come about by preventing events of a certain kind from happening (accidents and incidents) or actions of a certain type from taking place (violations and deviations). The decrease rather comes about because organisations learn to improve their everyday functioning and therefore succeed more often.

The development of RHC has been documented in the three volumes mentioned in Chapter 1 and in many journal papers – with McNab et al. (2016) as a recent example. The researchers and practitioners that have contributed to this development have through their work demonstrated how it can be done either by tailoring existing approaches and techniques to match a Safety-II perspective, or by developing new ones. Many of the approaches have involved the import of tried and tested methods from the social and behavioural sciences into health care settings. Since many of these methods are of a qualitative rather than a quantitative nature, it was initially somewhat at odds with the conventional clinical research methods that patient safety research had emulated (Vincent, Neale and Woloshynowych, 2001). This situation has fortunately slowly improved, although there is still some way to go.

Each of the many studies and projects that have been carried out has naturally focused on presenting the specific work and the specific outcomes. The motivation for providing a practical guide on how to deliver RHC such as this – if not actually a proper handbook of RHC, something still to come – is therefore to offer a helping hand to practitioners and researchers who are curious enough about RHC to want to try it in practice, but who may be uncertain about how to go about it.

The main part of this book comprises 13 chapters that each describe how RHC has been studied or applied. To make it easy for the readership, the editors have insisted that each chapter follows the same structure or layout. A possible chapter structure could have been the traditional layout of scientific papers with sections such as introduction, method, results and discussion. The editors, however, found that this was too generic to serve the purpose of the book. While it clearly is valuable that the chapters correspond to the accepted way of describing research and to common academic criteria of quality, it is equally important – at least – that the chapters also serve their pedagogical purpose. Rather than just present the usual Method section, for instance, chapters should also provide some detail about *why* a specific method was chosen and *what* the consequences were for, e.g., data sources and data collection. The main purpose of the chapters in this book is in short not just to present new findings, but also to describe how this was done. In order to achieve this each chapter addresses the following issues:

- The data collection methodology, which represents the *rationale* for the data collection approach. Because this often is done implicitly rather than explicitly, it is essential for the book that the chapters make the basis and arguments explicit.
- How *data sources* (e.g., observation opportunities, informants, data repositories) were selected or chosen. This includes both the practical concerns and constraints and the scientific criteria for the 'ideal' data for the case.
- How data were *collected*. Specifically issues of interest are how much time and how many resources were used, how organisational support was achieved, and more.
- How data were *analysed*. The purpose of the analysis is to make sense of the data, to aggregate data, to look for patterns and trends, and more. This is especially important for the more qualitative approaches that are used by RHC. In the common quantitative or statistical approaches, the assumptions that drive the data analyses are usually taken for granted

and are in most cases embedded in, and also hidden by, the specific methods.

- How the analysis results were *interpreted* and gave rise to conclusions, recommendations, and more.

In retrospect it is interesting to realise that none of the authors of the 13 chapters were asked to follow these guidelines when they planned and carried out their studies, but only when they wrote about them. That they nevertheless were able to do so without too much effort shows that these issues represent a prevailing feature of the scientific tradition – in health care, and in every other field as well. Yet when we do become used to things in a certain way we stop paying attention to the details. While this can be beneficial for something that is done routinely, because it can save time and effort, it is a potential limitation for something where thoroughness is more important than efficiency, such as trying to bring Resilient Health Care into practice (Hollnagel, 2009a).

Chapter 3

Procuring evidence for Resilient Health Care

Erik Hollnagel

Introduction

All sciences have their foundation in theories and models on the one hand and data and evidence on the other. In some cases the theories are well articulated and can guide the search for data, which then is seen as proof or corroboration of the theories. Examples of such exact or 'hard' sciences are mathematics, optics, astronomy and physics. The existence of dark matter and dark energy, for instance, is predicted by the standard model of cosmology, and the search is guided by the properties that can be derived from the theories. (Neither dark matter nor dark energy have, however, been found so far.)

In other cases models and theories are less well developed or articulated. This is characteristic of the so-called 'soft' sciences, or social sciences, which include sociology, management, economics, psychology and history. In the context of this book safety science and Resilience Engineering (RE) can confidently be added to the list, insofar as they can be considered sciences at all (Hollnagel, 2013b). For these sciences it can sometimes be difficult to produce testable predictions or to test predictions by means of controlled experiments, both of which – for better or worse – are seen as attractive qualities. There is therefore a need to base practice on data or evidence rather than on theories and models. This is not necessarily an impediment to research and development. Indeed, one positive consequence is that the findings are more likely to relate directly to, and therefore have a direct impact on, everyday experience.

Health care in general must also rely heavily on data or evidence, since theories often are vague or lacking. Evidence-based medicine is clearly an example of that, as the name implies. This is mainly because health care takes place in what we commonly refer to as a complex socio-technical system. Even if we leave out the adjective complex, the available theories of how organisations and socio-technical systems function and perform are generally insufficient to provide a consistent and systematic basis neither for daily practice, nor for research and development. Although there is no shortage of theories of how organisations function, few of them are detailed or specific enough to serve as the primary basis for day-to-day management. Another 'soft' science, economics, is perhaps an even more extreme example, to the extent that some theorists (sic!) question the very existence of economic theory (Ormerod, 1994).

Resilient Health Care (RHC) is definitely a young and possibly also a 'soft' science and is therefore much dependent on whatever data and experience can be made available. In order to make verifiable progress it is necessary that data and evidence are procured in an orderly and systematic manner. The main purpose of this book is to demonstrate how that can be done. There is fortunately a solid foundation for that, both from nearly two centuries' experience in social science and from the much longer experience with the scientific method, the time-honoured principles and practices for investigating phenomena and for managing knowledge – acquiring new knowledge as well as correcting and integrating previous knowledge. The classical example is the Baconian Method described in *Novum Organum* from 1620. This has become the standard scientific method, often presented as 'hypothesis–experiment–evaluation'. When this is rendered as 'plan-do-check', it is easily recognised as the foundation for the more recent 'plan-do-study-act', the PDSA or Deming cycle, that is one of the most widely used methods in health care management. Although the PDSA strictly speaking is a method for the continual improvement of processes and products, the similarity to the general scientific method is striking – and of course not fortuitous. When used in the context of research rather than 'just' quality improvement, it is nevertheless useful to pay heed to Deming's admonition that 'Experience by itself teaches nothing . . . Without theory, experience has no meaning. Without theory, one has no questions to ask. Hence without theory there is no learning' (Deming, 1994: 103).

A fundamental characteristic of the social sciences is that they do not deal with an independently constituted subject-matter that continues unperturbed regardless of what we do. In the natural sciences, scientists try to understand and theorise about the way the natural world is structured. The understanding goes in one direction only in the sense that what we try to understand remains inert and does not try to understand us. The situation is completely different in the social sciences where the understanding goes in both directions: that which we try to understand also tries to understand us, and changes in consequence of that. This is especially important to keep in mind when planning and implementing changes based on 'evidence' and is discussed further in Chapter 17.

To facilitate the description of how the studies described in this book were done, each chapter has tried as far as possible to use the same layout. The layout was chosen to bring forward the important considerations that always are part of research and development, but which often are neglected in the reports. The reason is that they, with not inconsiderable justification, can be taken for granted in the dominating research culture, hence be left unspoken of. Yet due to the lack of fully developed theories and models, as outlined in Chapter 2, it becomes a priority to be explicit in accounting for the details of the ways in which data were obtained. This is done by looking first at the reasons or rationale for the data collection approach, then at how the data sources were selected, then at the ways in which the actual data collection took place, then at how the data were analysed, and finally at how the outcomes of the data analysis were interpreted.

Rationale for studies and data collection approach

The general paradigm for scientific research, whether in the original form of the Baconian Method or the many later variations thereof, is the 'hypothesis–experiment–evaluation' triad. According to this tradition any kind of research and any kind of investigation must begin with a hypothesis or at least with some idea about how the phenomenon being studied can be understood – or more generally about 'how the world works'. The assumption that it is possible just to observe what happens in the real world is dangerously naive, since it disregards the fundamental fact that it is impossible to look for something unless there already is some kind of idea or preconception of what to look for. Just as there is no such thing as a blank mind, a *tabula rasa*, there is no such thing as an unframed observation.

Testing hypotheses

For sciences that are sufficiently mature, such as the natural or 'hard' sciences, the researcher or investigator can refer to a set of theories and models from which hypotheses can be produced. These will typically explain how causes and effects are related, hence enable predictions for which observations can be made or for which experiments can be designed so that the hypothesis can be verified or falsified. (One proposed distinction between the 'hard' and the 'soft' sciences is indeed that the former can produce hypotheses that can be falsified, while the latter cannot.) The rationale of hypotheses testing goes as follows. We predict from the theories that a given influence X will lead to a specific effect Y. In order to confirm this we therefore establish suitable circumstances that make it possible to introduce an influence X', which is an acceptable approximation that is functionally equivalent to X. We then observe what happens to determine whether the outcome is Y' (functionally equivalent to Y) or non-Y. This determination is rarely by direct observation but usually relies on quantities – sometimes large – of direct or indirect measures that can be analysed statistically.

Generating hypotheses

While testing hypotheses has been accepted as more or less the gold standard for scientific research and development, and in many cases also serves as an ideal for the 'soft' sciences, it is difficult to apply in the study of complex socio-technical systems. The reason is simply that it is impossible to describe let alone control all the factors and conditions that may influence how the system performs, hence to make testable predictions. It is impossible both because the life situations are inherently dynamic and intractable and because the presence of humans as parts of the system introduces variability that never can be controlled – the understanding is bidirectional and the system being studied is not inert and stable. The best one can do is to identify the most important factors, environmental, social and human, and collect sufficient data to gauge their actual influence. The aim should actually not be to have complete control of the situation since the variability not only is inevitable but also an important contribution to how the system performs.

Testing hypotheses should obviously be used in cases where it is reasonable to do so, but at the moment this is only possible for a small number of all problems or issues in Resilient Health Care. In cases where it is impossible to use experiments to test hypotheses, the alternative is to try to investigate the effects of specific changes or interventions. System changes are made in order to obtain a set of expected effects or consequences, often as a reaction to some kind of unwanted event or occurrence but also as part of efforts to improve system performance in one way or the other (safety, quality, effectiveness, and more). The more explicit the basis for the changes are, the more detailed the description of the consequences can be. The degree of detail in the description may well be insufficient to qualify as the test of a hypothesis but may be sufficient to determine the conditions needed to make specific observations, to identify the changes exactly, perhaps as the preparation for generating testable hypotheses at a later time.

The approach used to generate hypotheses is to characterise the expected effects sufficiently well to make specific and focused observations possible. We can use the assumptions about how the system works, perhaps as a quasi model, to characterise the set of effects Y that are the expected results of an influence X. We can then establish a situation where something takes place that corresponds to an influence, X', an approximation that is functionally equivalent to X. We study the outcomes to determine the effects, using the expected characteristics of Y as a guideline (i.e., as an aid to focus as well as a filter). Once the observed effect Y' has been characterised, we can use that to corroborate our assumptions about how the system works.

This approach is typical of experiments and investigations in the social sciences. In the case of humans (and social institutions), the complexity is often so great that it is impossible to enumerate the parameters. Instead large data sets are collected and analysed (often statistically) for recognisable relationships and patterns. In some cases these have been predicted, in others not. The main point of the exercise is that the relations can be explained *post hoc* by referring to the initial assumptions, although it means that they in some cases are statistical artefacts rather than empirical facts.

In this type of research, it may be more difficult to define reasonable and direct measures of performance. A larger proportion of the data may therefore come from various kinds of observations. These must naturally be prepared as well as possible, and observers must be experienced and adequately trained. The variability of human performance naturally limits the amount of details one can put into, for example, an observation schema, but the categories for observation should at least be so clearly defined that they do not jeopardise reliability.

Establishing empirical data

In some cases, particularly at an early stage of development of a science, the assumptions only suffice to support a general type of data gathering. They are ideas or hunches that point in certain directions and support focused observations but which cannot be articulated well enough to be the basis of a controlled or even semi-controlled study. In this kind of focused observations, the purpose is to find out what actually takes place in a given situation, possibly with the aim eventually to formulate hypotheses, if the purpose is scientific, or at least be able to plan actions and interventions if the purpose is practical. Rather than trying to study the effects of some sort of specific change, the aim is to take a closer look at a specific task or activity, specific subjects, and in general to look at Work-as-Done. The purpose may also be longitudinal as in trying to recognise predominant patterns in performance.

In these cases the observations are used to recognise patterns and to formulate the functional characteristics of the system. There is, of course, always the experimenter's understanding of what s/he is doing which provides the conditions for description. But it is through this type of experimentation and systematic observation that a better articulated description gradually is established. This approach to research is sometimes called hermeneutical, since the qualitative methods of analysis dominate. The goal is to develop the understanding of a domain or phenomenon, and refine the descriptions. This frequently happens in the early stages of a science (what some would call the pre-paradigmatic period). In some cases it might be profitable to emphasise breadth, i.e., to cover as many aspects as possible, and in others to emphasise depth, i.e., to go into the details of a single aspect. The decision of what approach to take must be made for each case in turn.

The characteristics of the three approaches outlined above are summarised in Figure 3.1. This can serve as a useful framework for taking a closer look at the main chapters of this book.

Different rationales for research into RHC

It is hardly surprising that none of the 13 chapters were testing specific hypotheses. Given the current state of development of RHC, that is rather

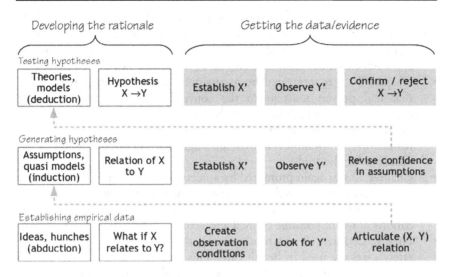

Figure 3.1 Three types of research rationale.

to be expected. The chapters rather fell into two sets of roughly the same size. One set comprised chapters where the rationale referred to accepted concepts or even quasi models in RHC. These studies match the category of generating hypotheses. The other set comprised chapters that basically can be characterised as focused studies or investigations of salient principles. These studies can therefore rightly be seen as serving to gather evidence.

One of the most influential ideas in resilience engineering and therefore also in RHC is the distinction between Work-as-Imagined (WAI) and Work-as-Done (WAD). It has even been the subject of the third book in the RHC series (Braithwaite, Wears and Hollnagel, 2017). WAI and WAD are common sense categories with overwhelming face validity, but do not as such refer to a specific theory of work or of how work goes well. (In contrast to that, there seem to be no shortage of theories about how things can go wrong.) Used as a research rationale it means that the focus is on developing descriptions of work as it actually takes place in real-life conditions, but possibly also on the more formal descriptions of work that is used in, e.g., management, or can be found in guidelines and procedures. A focus on WAD has clear implications for how a study should be planned, and for which types of data are interesting. It clearly also favours qualitative rather than quantitative data, hence *observations* of daily work rather than *measurements* of work outcomes.

- Chapter 5 studied the use of workarounds in everyday clinical work to get a better understanding of how and why things go well. It used a triangulated approach comprising ethnography with interviews and focus groups.

-

- Chapter 7 examined what strategies people used when they successfully followed procedures. In this case it was decided to use structured interviews rather than observations. The interviews were structured using an approach called the Resilience Markers Framework (RMF).
- Chapter 10 was motivated by a persistent problem with patient safety that had proven resistant to the conventional solutions. The study focused on the variability in daily routines as a way to understand normal work. The variability was captured using the Functional Resonance Analysis Method (FRAM), although that did not itself serve as the rationale for the study.
- Chapter 12 was again prompted by a safety problem, where the investigators recognised the need to understand how work was actually done. Experience had shown that there was a difference between the recommendations of the responsible council and actual practice.
- Finally, Chapter 15 was motivated by the need to look closer at how safety cases are used in health care. This was done by studying clinical handover in emergency care better to understand the gap between WAI and WAD.

Another distinct proposal from the resilience engineering and Safety-II is that resilient performance can be described as requiring four potentials: the potential to respond, the potential to monitor, the potential to learn, and the potential to anticipate. While it is not formally presented as a theory, it at least corresponds to a quasi model and perhaps even to a generic model of resilient performance (Hollnagel, 2017). It is also the basis for a specific method to develop and manage the resilience potentials known as the Resilience Analysis Grid (RAG). The notion of the four potentials was used in two chapters as the rationale for the approach and to help structure the observations and data collection.

- Chapter 8 applied the four potentials to explore the mechanisms shaping resilience in maternity services in the maternity wards of two Norwegian hospitals. It also compared the two hospitals to explore important patterns across different contexts.
- Chapter 11 went one step further and tried to refine the Resilient Assessment Grid so that it could be used by emergency care providers, ancillary staff and leaders to assess and monitor the potentials over time. It was therefore not only a hypothesis generating study but also a practical evaluation of an approach to the management of Resilient Health Care.

The rest of the chapters roughly fit the category of *investigations* in the sense that they described work undertaken to make a closer examination of something or systematically to look for specific kinds of evidence. In practice a study can, of course, serve several purposes at the same time, for instance making a broad sweep for relevant evidence or conducting a more focused search for data to improve the understanding of a certain phenomenon or relationship.

- Chapter 4 looked at how care of older people was coordinated and delivered, how decisions were made, and how people adapted to the pressured clinical environment. This was based on the assumption that resilient performance is an aspect of everyday clinical work and of how people make sense of their environment, hence this should be looked at more closely.
- Chapter 6 focused on the role of simulation to help understand and support the emergence of RHC. Based on previous simulation experiences it was assumed that important information could be uncovered via video analysis and guided participant debriefings using a mixed methods approach.
- Chapter 9 is a good example of a broad sweep for data. At the outset it was not clear what consequences the introduction of RHC principles in a Clinical Care Complex would have. The study therefore tried to get viewpoints from a variety of staff, primarily by means of interviews and in-practice observations.
- Chapter 13 took a close look at what happens in the clinical microsystem of in-patient diabetes care, focusing on the delivery of care as it works in practice, on how misaligned demands and resources affect daily work, and on how people adjusted their performance to that. The approach referred to a model developed by the Centre for Applied Resilience in Healthcare that also was used in Chapter 4.
- Chapter 14 used insights from RHC and RE to evaluate the consequences of conventional improvements to the preparation and administration of drugs. The study took place in the context of an 18-month long specialisation course in industrial engineering, and involved the 30 hospital employees who attended the course.
- Finally, Chapter 16 is a longitudinal study of how nursing performance was sustained in one of three acute medical wards that were relocated after the Christchurch earthquake in 2011. It looked specifically at shared leadership within the team and between charge nurse managers of all three relocated wards.

Choice of data source and data collection method

The rationale of a study defines the purpose – what is being studied – and also helps to determine how the study should be carried out. When the rationale has been determined the next question is what data will be needed and where they can be found. A related question is how the data can be collected or obtained. Since methods and techniques for data collection are closely related to the data sources, the two issues will be treated together here.

The choice of data sources and data collection method is relatively uncomplicated when the rationale is hypothesis testing. In this case the data sources are defined by the purpose, so to speak, and there is usually also enough knowledge about the system to determine how data can best be collected, measured or counted. A hypothesis will typically describe an expected dependency between independent and dependent variables, and the specification of the dependent variables is in effect a specification of data sources as well. If the rationale is to generate hypotheses, there is still some guidance to be found in the underlying assumptions, although the net must be cast wider than in the first case. There are no clearly defined dependent and independent variables, but there are assumptions and quasi relations between, e.g., conditions and a range of outcomes. If the rationale is to gather evidence and experience there may be even more uncertainty. The starting point will be a set of assumptions about how the phenomena or performances of interest happen, which means that there will be some knowledge about the system, and how it works. These assumptions can point to relevant data sources, though perhaps by exclusion as much as by inclusion.

Since the studies described in this book predominantly are of a qualitative nature, it follows that the data also mostly are qualitative. Most of the studies are of the inductive or abductive kind aiming to formulate theory or hypotheses, rather than of the deductive kind aimed at the testing of hypotheses, cf., Figure 3.1. This corresponds to differences in the type of data that are needed and in how they are treated or analysed, but the two types or schools of thinking are actually not fundamentally different. The differences between qualitative and quantitative research and data is a matter of degrees rather than absolutes. While many of the studies make use of what often is called 'subjective' data, i.e., the views, experiences and opinions of people who are directly involved in the work, some of them also include more 'objective' data, for instance from data bases or reports or observed effects (though through the interpretation or eyes of the researchers) of a program on a problem or condition, or reviews of documents. When the main focus is on how health care systems work, it is more useful to understand what goes on during work and how work is carried out than to count specific types of outcomes. Qualitative research involves more in-depth information on a few cases rather than less in-depth but more breadth of information across a large number of cases. The views of the people involved, as well as their performance as it can be observed directly or indirectly (recorded), therefore takes precedence over measures and magnitudes, i.e., data that can easily be quantified.

When getting the views of the people who have first-hand experience with the issues being studied, or when observing their performance, it is generally considered important to ensure that the informants have appropriate backgrounds and sufficient experience, i.e., that they are experts rather than novices. This does not mean that the experiences of someone novel to a situation cannot be valuable. Indeed they can. But they will be different from the experiences of someone who is familiar to the situation. The latter will also have developed the habits, for better or worse, that are common to the work, and which precisely are those that are being sought in studies of WAD.

Direct experiences

Direct access, where people directly express their experiences, is mainly through tried and tested methods such as interviews, individually or by groups, questionnaires and debriefings, either structured or unstructured. There is a copious literature on how to conduct interviews and develop questionnaires hence no need to go into it here. Especially in the case of interviews or discussions with several people at the same time there are well-established methods such as World Café, focus groups, process mapping sessions, etc. A special case is the use of autoethnography, a form of qualitative research in which authors use self-reflection and writing to explore their personal experience and connect this autobiographical story to wider cultural, political and social meanings and understandings.

Indirect experiences

Indirect access to experiences means that people are observed for a period of time instead of being asked directly to relate or express their experiences or views. The observations can be structured or partly controlled, as in the use of simulations, but they can also be more naturalistic and take place in the actual work environment, preferably without disturbing it too much. The observations can be purely subjective, supported by field notes, by video or audio recordings, etc. Such methods are often called ethnographic, and are derived from work in anthropology and sociology.

There are two important issues related to indirect experiences that are amply illustrated by chapters in this book. One has to do with the problem of 'going native'. The term means that researchers/observers become so familiar with the people being observed that they adopt the views and perspectives of the people being observed. This of course reduces the degree of objectivity that is seen as an indispensable quality of good research. The other issue has to do with what is called purposeful sampling. The opposite of purposeful sampling is random or stochastic sampling, where the most important criterion is that data selection is unbiased in the sense that each type of data has the same probability data of being chosen as any other. This is a central issue in quantitative research and in the testing of hypotheses. In qualitative research, and particularly in cases of hypothesis generation and experience gathering, it is more important to get data that are relevant for the purpose of the study. The criterion for the value of the data is not generalisability but credibility. The sampling is therefore guided by the purpose and the rationale. Several of the studies made an explicit note of that, but all studies did in fact use purposeful sampling by virtue of not being hypothesis testing. One criterion or guidance was the need to find out more about how work is done, about how people adjust their performance to the current and expected conditions, as this is one of the central themes of RHC.

In more than half of the studies the data sources were based on direct experiences (cf., above).

- Chapter 7 was based on interviews of six experienced anaesthetists who were alone with the interviewer. The interview followed the five-step structure of the Resilience Markers Framework. People were specifically invited to reflect on the different demands of their work and to contrast routine and non-routine aspects.
- Chapter 9 also relied on semi-structured interviews with qualified staff, in this case 24 nurses and four intensivists who had worked in the Critical Care Complex for at least two years. The purpose of the interviews was to find out people's experiences after the introduction of new ways of working, referred to as the Team Resilience Framework.
- Chapter 10 used structured interviews with experienced staff with different roles in the department: physiotherapist, chiropractor, medical secretary and registered nurse. Doctors were unable to participate due to understaffing. The interviews took place at the informant's workplace. The interviews referred to a specific interview guide in order to elicit information about the functions that are part of everyday work and how these can be characterised.
- Chapter 11 secured data from people in groups, using a technique called the World Café (Brown and Isaacs, 2005). The sampling here was also purposive, but the criterion was to get a broad range of perspectives and voices. Participants were chosen from within the department as well as from outside, and included relative new-comers as well as very experienced people. They were invited to reflect on and share their overall perceptions of 'resilience' in the Emergency Department, and how their work environment helps or hinders them in providing safe care. The dialogue workshops were conducted in a near-by off-site meeting room during the partici-pants' non-working hours.
- Chapter 13 again chose in-depth interviews but in a slightly different way. The interviewees all had relevant experience, although they did not need to be specialists. The interviews were conducted on a

one-to-one basis until data saturation had been achieved and range of staff at all levels had been consulted. The main focus was information about the routine adjustments necessary for care.

- Chapter 16 used direct data but in a rather different way, namely by means of autoethnography. This means that the author relied on self-reflection and writing to explore his personal experience while connecting this to a wider context. This involved two steps, creating descriptive narrative text and then gathering of reflective notes. The focus was guided by the four potentials for resilient performance described previously.

The remaining chapters relied mainly on indirect experiences, in the sense that data included what people did in addition to what they said. The data in nearly all cases also included more objective or depersonalised data such as documents, reports, etc.

- Chapter 4 used researchers to observe the activities of the clinical staff. The observers were at first stationed themselves in the centre of each ward but gradually changed to 'shadowing' – i.e., closely following – clinical staff while they were performing their daily work. The focus of observations was also gradually refined as familiarity with the setting increased and relationships with staff developed. Interviews were also conducted with a range of staff from different disciplines and levels of seniority. The observations were documented by extensive field notes, which later were transcribed and expanded following the observation session and saved using archiving software.
- Chapter 5 made use of extensive observations of people who were experienced with the medication management system at two different hospitals but covering entire shifts rather than just the medication administration process. The study used a triangulated approach to data collection consisting of observations (non-shadow and shadow), interviews and focus groups. This was helpful to provide different perspectives of nurses' enactment and explanation of workarounds.
- Chapter 6 was based on data obtained from medical simulations. This comprised video recordings, but the most important source was guided debriefings that followed directly after each simulation session. The interest centred on the consequences of changes in team configuration, resources and facility layout. Additional data were obtained by using the NASA-TLX, a widely used multidimensional tool to assess perceived workload.
- Chapter 8 made use of qualitative interviews, focus group interviews, field notes from observations and analysis of national documents. The interview guides focused on a diverse range of issues such as role, responsibilities, teamwork, culture, management, quality improvement,

IT systems and how the participants experience work at their unit. The data sets were extensive, covering 99 interviews, field notes from 45 hours of observation, and focus group interviews collected during 2011–2012.

- Chapter 12 used various data sources that directly or indirectly represented WAD. This included data from the researchers' own hospital as well as from other hospitals, minutes and memoranda of hospital committees, medication supply data, observations of medication practices by ward pharmacists and interviews with clinicians. External data sources included expert opinions of critical care physicians by means of an open mailing list and interviews with patient safety officers and managers of other hospitals with similar characteristics.

- Chapter 14 also used a broad range of data sources selected to include multiple sources of evidence, a mix of quantitative and qualitative data, perspectives of different agents involved in the socio-technical system, formal (theory-based) system descriptions and an understanding of work-as-done. This comprised participant observation, time and motion studies, documents and focus groups. Data collection took place over a period of three months.

- Chapter 15 employed a different approach based on half-day workshops or sessions with stakeholders from different departments and services in the hospital. Observations also took place for several hours a week in the emergency department for a period of seven months. And, finally, this study included semi-structured interviews with participants from different parts of the services.

Analysis of data

It is a paradox of research that the better structured and the more systematic or rigorous the approach or research paradigm is, the less visible are the underlying assumptions. Researchers that rely on standardised or textbook methods usually apply them without thinking too much about what the methods actually do and why. But even highly standardised calculation methods that are the core of statistics rely on a number of assumptions, as recognised already by Fisher (1926). The blank acceptance of standard methods is, of course, justified because they have been proven through many years and many uses. Standard methods represent the social consensus – if not the social construct – that a science has of how the world works. A possible unwanted consequence, however, is that researchers when faced with a new problem may fall into the trap of using the familiar methods without considering whether they are appropriate, i.e., whether the assumptions hidden in the methods actually are reasonable under different circumstances. Too many researchers in both the science and social science traditions are methods-driven.

In the case of hypothesis testing, the data analysis serves a clear purpose, namely to determine whether a hypothesis should be rejected or verified. In this case there is a number of well-defined methods and rigorous criteria, or rules, for when something is acceptable or not, especially in the case of numerical data and statistical analyses.

In the case of hypothesis generation, the data analysis serves the purpose of identifying (or proposing) patterns in the data, that in turn can become quasi theories and eventually theories. (In many cases they are heralded as theories or models already when they are first pronounced even though they are neither. This is because both terms frequently are used rather imprecisely, especially the term 'model'.)

In the case of gathering evidence, the analysis is even more in the nature of being an interpretation or of making sense of the data. This corresponds to looking for and recognising patterns, which then in the case of hypothesis generation can be generalised even further to become proto-hypotheses. The patterns are recognised in the sense that they are imposed or proposed as an interpretation or explanation of a phenomenon or performance. It is obviously important that the patterns are intersubjectively accepted, i.e., that many people can agree that a proposed pattern is a sensible way of categorising and thereby interpreting the data.

In all types of research the data can be seen as going through several stages during the analysis. The presence of such stages is, however, easier to see, the less standardised the approach is. Indeed, the lack of standardisation also means that there is a greater need to be explicit about how the data analysis takes place. The chapters of this book provide excellent examples of that. The general steps in data analysis are useful as a basis for understanding how the analysis takes place. Regardless of the purpose of data collection (hypothesis, testing, hypothesis generation, or the gathering of evidence), it is possible to distinguish between five stages in the refinement (processing or abstraction) of data. The different stages illustrate the dependence between data collection and data analysis. It is important to acknowledge this common conceptual foundation because it provides a basis for selecting a proper data analysis method and the possibility for cross-checking results.

The analysis of data on human performance may conveniently be described by the following five stages:

• Raw data. The raw data constitute the basis from which an analysis is made. Raw data can be regarded as the fragments of a phenomenon or performance, in the sense that the raw data do not themselves provide a coherent description but rather serve as the necessary building blocks or fragments for such a description. Raw data constitutes the elementary level of data for a given set of conditions. The level of raw data may thus vary from system to system and from situation to situation.

- Intermediate data format. This format represents the outcomes of the initial processing of the raw data, for instance by combining or ordering them in some way. In many cases this makes use of a time line, to provide a coherent account of a phenomenon or performance. It is thus a description of the actual performance but in the terms of a professional user (domain expert) rather than of a researcher (analysis expert), hence the language from the raw data level rather than a refined, theoretically oriented language. The step from raw data to the intermediate data format is relatively simple since it is more a rearrangement than an interpretation of the raw data.

- Categorised event data. At this stage the intermediate data format has been transformed into a description of the task or performance using terms and concepts that reflect the theoretical basis for the analysis. Performance descriptions may still be ordered along a time line which is specific to the situation in question but uses the categories related to important concepts or assumptions. The transformation changes the description of the actual performance to a formal description of the performance during the observed event. The step from the intermediate data format to the categorised data may be quite elaborate, since it essentially is a theoretical analysis of the actual performance. The emphasis is also changed from providing a description of what happened to providing an explanation as well.

- Conceptual description. At this stage of the analysis the description no longer refers to a particular situation but is rather aimed at presenting the common features from a number of situations. Describing what is common to multiple performances leads to a description of the generic or prototypical performance. This may still be ordered along some time line but the underlying dimension is released from the specific events. The step from the formal to the prototypical performance is typically quite elaborate and requires an analyst with considerable experience, in addition to various specialised translation aids.

- Competence description. The final stage of the data analysis connects the conceptual description to the theoretical background. The competence description is concerned with the basic concepts or ideas, which in the case of RHC (and resilience engineering) are the potentials for resilient performance, WAD, performance variability, workarounds, trade-offs, etc. The description of competence should be relatively context independent: it is the description of the behavioural repertoire of people and organisations independent of any particular situation though, of course, still restricted to certain classes of situations. As soon as a context is provided the description of competence can become a description of prototypical performance and, pending further information and specification, a description of the typical performance. The competence description can thus be seen as the basis for performance prediction, for instance during system design.

The relations between the five stages can be rendered as shown in Figure 3.2. The rendering uses the idea of functions, and illustrates the relations between the stages (levels of processing of data) and the characteristic functions in data analysis.

All the studies described in this book focus on the 'third' function, namely the identification of patterns and typical traits. This can also be described as a thematic analysis, which is a common analysis form in qualitative research. As the name implies, thematic analysis looks for recognisable themes or patterns in the data that point to generic (nomothetic) issues, things that are common to phenomena and performances as they are seen in everyday performance. The development of Resilient Health Care, as well as of resilience engineering, has emphasised or identified several salient phenomena. Prime among these is the nature of everyday performance (WAD), specifically the intentional use of performance adjustments and workarounds to overcome concrete problems. Another example is the quasi-theory, if such a term is allowed, about the four potentials for resilient performance, either in the common description or as part of the RAG. The chapters also provide other examples of patterns that were found useful for the specific studies. It is fair to say that all studies made use of thematic analysis in one way or another, as seen by the brief account of each chapter in the following:

- Chapter 4 analysed the data in several phases or steps. First a set of resilience narratives was written for each six themes. The resilience narratives were extended descriptions of how the unit functions in terms of features that are important for resilience. The narratives were then reviewed and discussed by the team and amended to reflect an updated understanding of the data.
- Chapter 5 used a general inductive approach to characterise key themes and processes (workarounds and performance adjustments) in the data in combination with a deductive analysis. Together these served to frame the key themes and processes against the research questions and to highlight the importance of identifying the goal of the workaround.
- Chapter 6, the analysis of the simulations by means of information obtained through debriefing identified several themes in team performance as well as 37 specific latent safety threats assigned to four different

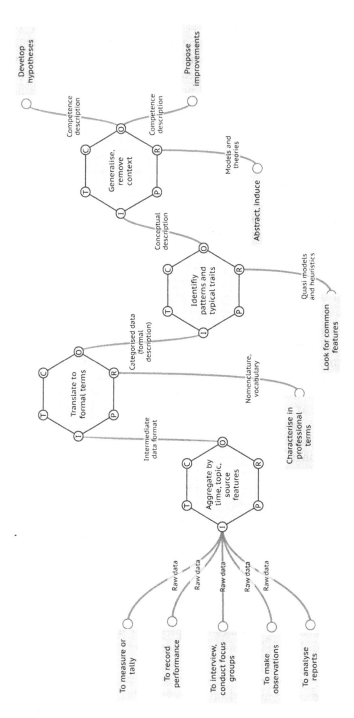

Figure 3.2 Stages in data analysis as coupled functions.

categories. The simulation performance was also analysed from the perspective of the four potentials for resilient performance.

- Chapter 7 analysed the results from the semi-structured interviews looking for specific examples of resilient behaviour. It looked for vulnerabilities and how people responded to them e.g., by modifying their behaviour.
- Chapter 8 performed a content analysis of the interviews, where the text was sorted using the four potentials for resilient performance as the conceptual structure. The raw data came from two different hospitals (micro-systems). The analyses on the micro-system level were then compared in a cross-case to look for factors that promote response, monitoring, learning and anticipation, respectively.
- Chapter 9 analysed the interviews using an inductive thematic analysis to highlight interesting phrases and recurrent terms or ideas. This went through several iterations by means of which data and categories were reviewed and collapsed until larger, overarching themes could be identified.
- Chapter 10 did a thematic analysis of the data using the principles of the Functional Resonance Analysis Method (FRAM) to provide the necessary structure. This was used to look for common patterns in how work was performed.
- Chapter 11 took the transcripts from the group discussions as whole units of discourse. These were together with field notes read and re-read to bring out meanings, patterns and themes by gradually refining emergent patterns. The outcomes were compared and contrasted to look for discrepancies, coherence and complementarity – all with reference to the four potentials and the RAG.
- Chapter 12 had seven different types of raw data from the researchers own hospital, from other hospitals, and nationwide. The analysis tried to determine why the recommendations from the authorities were not followed, and paid special attention to the differences between WAI, as the basis for the recommendations, and WAD.
- Chapter 13 analysed transcripts of interviews using an iterative qualitative analysis technique. The reference was the same model as in Chapter 4. The analysis looked for high level concepts of demands, capacities, adjustments and outcomes. This was in turn used as a basis for identifying so-called resilience trajectories.
- Chapter 14 used content analysis of notes from meetings, observation diaries and other documents as the main method. The analysis was guided by the principles underlying Value Stream Mapping and FRAM, respectively. This was supplemented by the analysis of data from time and motion studies.

- Chapter 15 used traditional risk analyses as a basis for considering how functions could vary and how the variability could spread through the system. This was then used together with the participants' reflections, the observational data and the interview results to look for characteristic trade-offs.
- Chapter 16 used a deductive qualitative analysis of the autoethnographical data. The reference for the analysis was the four potentials for resilient performance.

Interpretation of data

The penultimate step of all studies is the interpretation of the outcomes of the data analysis. If the study serves a practical purpose, a further step may be the development of recommendations for appropriate actions, changes or interventions. The findings provide the knowledge that is a prerequisite for the potential actions.

The purpose of this section is, however, not to summarise the findings from the various studies in terms of their contents, but rather to look at how the interpretations typically were made, i.e., the principles and approaches that can be found in the various studies. The purpose of this book is to provide guidance on delivering RHC where the individual chapters present both the methods used and the actual findings. Although the chapters do not cover all the ways in which research data can be analysed and interpreted, they are fairly representative of the efforts to improve the knowledge and practices of Resilient Health Care at the time of writing.

The interpretation of the analysed data should take place within the context of the overall study to avoid the risk of extrapolating from single instances that may be inaccurate representations of the phenomena or performances in focus. An interpretation is literally an explanation of what something means given a specific context. An interpretation will always be relative to some conceptual background, whether formulated loosely as common sense understanding or more rigorously as a theory or a model. In the latter case the interpretation can take the form of the confirmation or rejection of a hypothesis. When the purpose of the study is rather to generate hypotheses or establish empirical evidence, the interpretation may serve to confirm and strengthen the relevance of studying the phenomena or performances of interest. (It would, of course, also be possible to reject, but it is rarely done.) Another purposes may be to provide support for a specific approach or methodology in which case the interpretation can show the value in doing things in a certain way, using a specific framework or tool.

The 13 chapters that follow illustrate both types. One type concerns a specific method or approach, in the sense of a framework or quasi model

that guides and supports the interpretation. The other concerns the usefulness of looking at things in a certain way, in the sense that it makes it easier
to make sense of the performance. In both cases, using a common reference
or shared understanding will of course also make communication easier and
more effective.

Most of the chapters focused on the benefits of taking a fresh look at the
problems found in health care using the RHC perspective. This could
either be as a closer study and improved understanding of WAD, or as the
use of a more specific view such as promoted by the RAG (the four
potentials) or FRAM (functions and performance variability).

- Chapter 5 looked at the relationship between workarounds and purposes by looking at how nurses adjust their work to competing
  demands under expected or unexpected conditions in order to sustain
  required performance. The study identified workarounds linked to five
  specific purposes, and used this understanding of WAD to argue that
  more attention should be paid to the potential impact of introducing
  new technologies and policies.
- Chapter 8 also argued for the value of having a better understanding of
  WAD. The data were analysed using the framework of the four
  potentials for resilient performance and this was helpful to identify
  common 'mechanisms' that people use in everyday work.
- Chapter 9 had the more explicit purpose of trying to increase the
  resilience capacity of the Critical Care Complex by focusing on the
  potentials to respond, monitor, learn and anticipate. This was done
  relative to a framework embodying six principles for effective team
  performance. The study showed how the data from interviews and
  observations could be interpreted to provide a coherent understanding of how the changes made contributed to increasing resilient
  performance.
- Chapter 11 also looked at an intervention that took place over a period
  of time, described as an organisational learning journey based on the
  RAG. The structure provided by the RAG and the four potentials was
  useful both in guiding the group sessions and in managing and
  interpreting the developments.
- Chapter 12 looked at a specific problem from several sides, and there were
  therefore several outcomes to consider. Taken together, the various
  findings agreed on the value of looking at WAD to find recurrent patterns
  in the responses to external pressures. In addition to that, the study also
  considered the risks that may be a consequence of the adjustments in
  WAD. Considering these findings together, the study argued for the need
  of a systemic approach.
- Chapter 13 looked at clinical care through using the elements of the
  CARE model to guide the interpretation. As well as demonstrating

the value of looking closely at WAD, the interpretation of the data pointed to the misalignment between demands and to the usefulness of identifying so-called 'resilience trajectories' or the ways in which clinical staff adjust and organise their work to be able to perform as needed.

- Chapter 14 interpreted the findings from two different perspectives, namely complexity and resilience. Doing so increases the confidence in the conclusions, which were expressed as six specific recommendations. The interpretation was extended by an initial analysis of the impact of the proposed countermeasures, thereby bringing a wider perspective to the study.
- Chapter 15 aimed at an improved understanding of WAD, specifically performance variability and dynamic trade-offs, in order to be able to understand better the value of safety cases in health care. The interpretation also included a systematic identification of risks beyond what is usually done, and led to the conclusion that the safety case approach cannot be used in health care without being adapted to the characteristics of this field of activity.
- Chapter 16 differed from the other chapters because the data were the result of using an autoethnographic method. The interpretation was guided by the notion of the four potentials for resilient performance, and the data were scrutinised from the four perspectives. This was then used as a basis for considerations about the role of specific forms of leadership in crisis situations.

The remaining chapters used the interpretation to assess the value in doing things in a certain way, i.e., a specific technique or approach, or looking at things using a specific way of structuring the data.

- Chapter 4 used the model developed in the CARE project to guide the analysis and interpretation (see Chapter 4 for details). The data (narratives) were coded using a number of central themes in RHC and analysed to identify opportunities for improvement. This in turn led to a proposal for devising ways to monitor specific aspects of performance, accepting that performance adjustments are inevitable because they are useful.
- Chapter 6 as a whole looked at the use of simulation. From looking at the overall results, one finding was that simulation can provide data and information that are unattainable from direct observations of WAD. The study also looked at the data from the point of view of the four potentials; the findings pointed both to a potential risk but also to ways in which the potentials for resilient performance in critical situations could be improved.
- Chapter 7 used an in-house framework called the Resilience Markers Framework, originally developed for a different domain, to explore the

interview data and identify resilience strategies. Next, the identified strategies were classified in terms of four themes representing perceived vulnerabilities and opportunities the in anaesthetists' work. The data also showed that strategies could differ between hospitals.

• Chapter 10 used the four fundamental principles of the FRAM to interpret the data. This was useful to clarify how work conditions and variability had an impact on everyday performance and pointed to three points of specific interest. The interpretation was also the basis for proposals of how to improve conditions. A further finding was that participation in the project contributed to changing the outlook of risk managers and clinical staff.

Conclusion

Resilient Health Care is still a young discipline, but it is making rapid progress driven by considerable enthusiasm. It was proposed as an alternative perspective to the established understanding of patient safety, in much the same way that resilience engineering was proposed as an alternative perspective to the established understanding of industrial safety. As the three previous books in this series have documented, RHC has in a relatively short period of time been able to finds its legs and to go from being a perspective to become a practice. In order to continue and strengthen this development, it is necessary to gain more directly applicable experience, to develop effective approaches and methods, and ultimately to be able to establish a basis of articulated theories and models.

The 13 chapters that follow each present a practical example of how the RHC perspective has been turned into practice. The chapters, however, not only report the results but also go into the details of how the studies took place. This chapter has tried to provide an overview or a summary of that by highlighting the important questions that must be asked by any scientific or practical study, namely: (1) what is the reason or purpose for the study, (2) what are the required data sources, (3) how can the data be collected, (4) how can the data be analysed, and finally (5) how can the results of the analysis be interpreted? Keeping these five questions in mind will hopefully both make it easier to read the chapters, and also help readers in finding ways to apply RHC in their own daily practice.

Chapter 4

Resilience Engineering for quality improvement

Case study in a unit for the care of older people

Janet E. Anderson, Alistair Ross, Jonathan Back, Myanna Duncan, Adrian Hopper, Patricia Snell and Peter Jaye

Introduction

Despite increasing interest in the principles of Resilience Engineering (RE) there is little guidance available for applying the ideas in practice in health care (Anderson et al., 2013). In this chapter we report on a study of resilience in the Older Persons' Unit (OPU) of a large London teaching hospital in which we developed practical tools to study resilience and identify potential quality improvement initiatives. This study is part of the work of the Centre for Applied Resilience in Healthcare (CARe), which also includes an Emergency Department as a study site. In this chapter we report initial results from the OPU site to illustrate how we have used RE principles to inform quality improvement. This study is unusual in that it was deliberately framed as a quality improvement initiative. Many other studies applying RE principles provide insights into how work is achieved and how resilience is manifest in clinical work (for example, Sujan, Spurgeon and Cook, 2015b; Nemeth et al., 2006), but in this study a further aim was to understand how we could use these insights to improve quality. Over the past few years discussion with clinicians, managers and policy makers about RE principles has found them very receptive to the ideas but uncertain about how to apply them in practice. Our view is that unless RE principles can be shown to improve the quality of care, with clear guidance on their practical application, the ideas will not be taken up in practice. We expect the results of the overall study (by early 2017) to enable us to extend and develop the theoretical framework of RE based on practical experience of implementing a resilience approach to quality improvement.

The aims of the study were to investigate organisational processes from a resilience perspective and on the basis of the findings design and implement interventions to increase the quality and safety of care. A resilience perspective was defined in this study as a focus on key RE theoretical concepts – variability in the environment, adaptations performed to cope with variability, goal trade-offs and variability in outcomes (Hollnagel, Wears and Braithwaite, 2015). In particular, we wanted to understand how resilience enables adaptive responses and affects outcomes. Resilience can be defined as 'the ability of the health care system (a clinic, a ward, a hospital, a country) to adjust its functioning prior to, during, or following events (changes, disturbances and opportunities), and thereby sustain required operations under both expected and unexpected conditions' (Wears, Hollnagel and Braithwaite, 2015a, p. xvii). It is important to note that this definition emphasises that resilience is an ability or capacity for adaptation, rather than a state of the system. Adaptive capacity is thought to be underpinned by four abilities that enable resilient performance – responding to problems, monitoring the state of the work system, anticipating future developments and learning from previous experiences (Hollnagel, 2009b). Activities related to these four abilities are observable, for example, responses to pressures and problems can be observed, as can learning mechanisms such as case reviews.

The first phase of the study was an in depth exploration of organisational performance using an RE lens. This required an operational definition of resilience concepts that could be used to guide the empirical work. The Concepts for Applying Resilience Engineering (CARE) theoretical model was developed and used to design data collection instruments, analysis methods and interpretation of the data. More detail about the model can be found in Anderson, Ross and Jaye (2016) and Anderson et al. (2016). Briefly, the CARE model, shown in Figure 4.1, posits that misalignments between demand and capacity create variability in the environment that drives adaptive responses by clinicians who strive to continue delivering care despite anomalies and disturbances. Work-as-Imagined (WAI) is represented in the model as the supposed alignment between demand

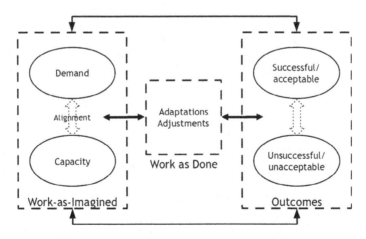

Figure 4.1 Concepts for applying Resilience Engineering (CARE) Model.

and capacity, and Work-as-Done (WAD) occurs as clinicians adapt in a fluid process of adjustment to misalignments such as missing staff or equipment. Outcomes, both successes and failures, emerge from the complex interplay between misalignments and clinicians' adaptations. Recursive relationships between all elements in the model capture the fluidity of an environment where outcomes can create future demands and capacities. For example, adverse outcomes can lead to the imposition of performance targets, creating a further demand on the system.

From the model the phenomena of interest were identified as misalignments, adaptations and outcomes. Concepts from the RE literature were also identified: goal trade-offs, learning from what goes right and the four resilience abilities of responding, monitoring, anticipating and learning. This conceptual work formed the basis of the approach taken in this study. In the following sections we describe how we designed data collection and analysis procedures to address these problems.

Methodology

Methodology refers to how researchers explain and justify the choices they make about all aspects of the research, including what data were collected, why and how, and how they were interpreted. These choices are shaped by various factors, including the researchers' views of the nature of the phenomenon being studied (ontology), and therefore what counts as knowledge about it (epistemology) (Carter and Little, 2007; Starks and Trinidad, 2007). Ontological and epistemological positions are often held implicitly rather than being explicated in research reports, but it is important for RE researchers to be explicit. The potential for misunderstanding and confusion is large in a new, multidisciplinary field of study, especially as the term resilience is used in multiple ways in different disciplines. RE research is at an early stage of development and clarity about the philosophical foundations of research projects can only help in understanding and defining the field.

Ontologically, our understanding of resilience was initially based on a definition that refers to the capacity of an organisation or unit to adapt to pressures and problems (Wears, Hollnagel and Braithwaite, 2015a). We reasoned that in a complex organisation adaptive capacity is likely to be distributed in complex ways through different health care functions and activities, making it difficult to observe directly. It is not codified or describable in terms of any one person, role, process or unit, but resilient performance is likely to be observable when good outcomes are achieved despite pressures and difficulties. Our ontological perspective of the nature of resilience suggested that studying resilience would require a subtle, nuanced and in depth understanding of the clinical environment.

This ontological stance shaped our view of what data would be of value in shedding light on resilience or, in other words, our epistemology. RE theory could not assist very much as it does not provide detail about its practical application and there have been few other studies that could provide guidance about application in a clinical setting with the specific aim of improving quality. Therefore we reasoned that as resilient performance was an aspect of everyday clinical work, it would be important to focus on how care was co-ordinated and delivered, how decisions were made, and how people adapted to the pressured clinical environment. Studying resilience, for us, would entail studying the social world of the clinic, how the actors make sense of their environment and take action on the basis of their understanding. Thus we used a broadly interpretive approach to understand the clinical world using participants' understanding of their environment and our own understanding of resilience (Ritchie et al., 2013).

This interpretive approach was supplemented by a combined inductive/deductive strategy for data collection and analysis. The adoption, a priori, of a framework of broad RE concepts was necessary to interpret the activities of everyday clinical work and generate insights into the resilience of the system. Deduction was therefore combined with induction at data collection, analysis and interpretation phases. Although some accounts of scientific processes view induction and deduction as mutually exclusive, there is a growing awareness of the need to employ both in seeking to understand complex phenomena (for example, see Langley, 1999; Wallace et al., 2003). In this study, the research proceeded by moving in a dialectical process between induction from the data and deduction from RE theoretical concepts. To support this process we used a formative strategy to gradually refine the focus of the fieldwork as familiarity with the settings developed and researchers identified ambiguities or subtleties that needed further investigation. The use of combined induction/deduction and the development of understanding through repeated cycles of data collection and analysis is analogous to Ricoeur's hermeneutic arc in which understanding is first developed by induction from experience or expectation, then tested objectively before resulting in a new understanding (Thompson, 1981).

Given our understanding of the nature of resilience (ontological stance) and our interpretivist epistemological position (supplemented by a combined inductive/deductive approach), immersion in the clinical environment, close observation of clinical work and interviews with practitioners were the best methods to use. The approach used has much in common with ethnography. Although the nature of ethnography has recently been debated, it is a research tradition in which immersion in the study setting and understanding through observations and interviews is emphasised (Jowsey, 2016; Dixon-Woods and Shojania, 2016).

Settings

The study was conducted in the acute medicine directorate in a large London teaching hospital. The OPU consists of three wards with an overarching management structure and specialises in the care of older people with chronic conditions and multiple co-morbidities, most of whom are admitted via the hospitals' Emergency Department. The OPU is staffed with a multidisciplinary team of doctors, nurses, occupational therapists, physiotherapists and health care assistants who deliver most of the personal nursing care. Fieldwork was carried out in all three wards of the unit.

Data sources

Ethical approval for the exploratory stage of the study was obtained from the University ethics committee. It initially covered only the first phase of observations and interviews, with the requirement to submit a further application when the details of the interventions were known. The research began with general observations of the organisation of the OPU and its structure and functions. Researchers initially stationed themselves in the centre of each ward from where they could observe activities. After familiarisation they shadowed clinical staff while they were performing their daily work and interacted with them naturally as the opportunity arose. In all cases care was taken to avoid encroaching on intimate procedures or witnessing confidential exchanges between staff and patients. In keeping with the formative approach, the focus of observations was gradually refined as familiarity with the setting increased and relationships with staff developed. During the early stages of data collection, key co-ordinating mechanisms within the OPU were identified, including multidisciplinary team meetings, board rounds, ward rounds, staff meetings and handovers. Observation was then expanded to include these activities. By maintaining a constant presence in the department, researchers became well known to staff and patients and were able to develop a high level of familiarity with the clinical environment. In total, 50 hours of formal observation were conducted. Interviews were also conducted with a range of staff from different disciplines and levels of seniority in which issues identified in observations or from the RE literature could be explored and clarified. Fourteen semi structured interviews were conducted.

Data collection

Existing research relationships between members of the research team and key personnel in the OPU were crucial in facilitating access to the study sites. The research team includes clinicians from the study settings, and this was essential in enabling researchers to access the settings and also for raising awareness of the study and gaining support from other key stakeholders. Clinical members of the research team also assisted with understanding the data collected and refining the focus of the study. In the course of data collection, and at formal forums, the research team has had contact with senior members of the hospital and as a result the study has a high profile within the organisation. This means there is support for the aims of the study and the potential interventions, which will be helpful in implementation.

Two researchers (MD and JB) working full time collected all the data. Extensive field notes, including reflections on the experiences of the researchers, were made by hand during observations. These were then transcribed and expanded following the observation session and saved electronically using archiving software. Semi structured interviews were conducted with a topic guide and were recorded electronically. Interview data were transcribed professionally.

Data analysis

The study generated a large amount of qualitative data and the challenge was to identify a way to capture the important features of organisational resilience and organise it in a coherent and meaningful way. The analysis was an extended multi-phase process as described below.

Identification of themes

The aim of the first step was to identify the main features of resilience in each setting using an inductive data-driven process. The extended research team met many times to reflect on and discuss the raw data and become familiar with it. A coarse-grained thematic analysis was conducted to identify the main features of the data. A tentative thematic structure was identified through discussion and this was tested with the data by using the search function in the analysis software to retrieve relevant data (for example, searching for the term discharge to retrieve relevant data on discharge planning). The research team then reviewed the data retrieved for each theme to determine whether the data were coherent and that the theme name aptly summarised the data. This process was repeated several times until the themes identified were coherent and informative. Six broad themes were identified that captured important aspects of work in the OPU. They were: performance review and learning, discharge planning and multidisciplinary team meetings, family and social services, staff roles and clinical task co-ordination, care delivery, and acute disruptions.

Resilience narratives

A resilience narrative was written for each of the six themes. Narrative methods of data analysis and reporting are often used when the aim is to make sense of a complex process. Narrative accounts have a story telling structure which is chronological, but non-linear. They describe trajectories of action (Greenhalgh, Russell and Swinglehurst, 2005) embedded in the social and organisational context within which they occur. Events and actions are linked to explain and make sense of actions. The resilience narratives are extended descriptions of the data for each theme, answering the questions – how is work organised (for example, in relation to discharge or interacting with family), what varies, why, and how are outcomes achieved? Writing the narratives required some interpretation but the accounts were kept as close to the data as possible. As a whole, the resilience narratives provided an extended description of how the OPU functions, organised by the features that are important for resilience.

Review and validation

The narratives were then reviewed by the extended research team including clinicians with knowledge of the OPU and RE principles (AH). Reviewers were instructed to review the narratives for accuracy and omissions and to provide alternative or supplementary interpretations if appropriate. After discussion, the narratives were then amended to reflect the updated understanding of the data.

Mapping to theory

The aim of this step in the analysis was to link the data to RE concepts using a deductive approach. The resilience narratives were coded using the following themes derived from RE theory.

- Misalignments between demand and capacity or pressures.
- Variability – what varies and why?
- Adaptations – how do people respond to pressures and variability?
- Goal trade-offs and prioritising decisions.
- Anticipation – how are future developments planned for?
- Monitoring – how is the system status monitored and problems identified?
- Responding – who responds to problems and what do they do?
- Learning – how is previous experience captured, analysed and used for learning?

Extended summaries were then produced describing the theoretical themes for each narrative.

Identifying quality improvement interventions

Finally, the narratives coded with RE theoretical concepts were analysed to identify opportunities for improvement. This was an interpretive process involving discussion with the extended research team and a clinical advisory group.

Interpretation

In this section we focus on discharge planning and provide an example of how data were interpreted to support one of the quality improvement interventions that was proposed as a result of this study. The purpose is to illustrate the logical link between the data collected, the analysis process and the final proposed intervention. Although we focus on a subset of the data for the purposes of illustration, it is important to note that during the analysis process isolated vignettes were not taken out of context or considered alone. Data were always analysed within the context of the overall study findings to avoid the risk of extrapolating from single instances that did not represent the work setting accurately.

Discharge planning occurred at regular multidisciplinary team (MDT) meetings. Formal consultant-led MDT meetings were held regularly on week days to bring staff groups together to discuss individual patients and plan timely and safe discharge. The meetings provided a forum for staff to discuss patient care and develop a shared understanding of the goals and progress for each patient. Each disciplinary group contributed insights that were not available to the rest of the team. For example, physiotherapists contributed insight into mobility problems and occupational therapists had information about social care needs. Staff members were expected to prepare for and attend these meetings and tasks were allocated at the meeting for each staff group.

From the observational field notes it was clear that planning discharge was difficult and time consuming due to the need to co-ordinate many different priorities and resources and the changing physiological status of the patient. The following excerpts from the field notes illustrate some of the challenges.

> *Query about why a particular patient did not go home as planned. Talking amongst physiotherapy team and junior Doctors. Estimated discharge date (EDD) modified on whiteboard.*
>
> *Patient discharged but no family at home, this causes stress for a staff nurse who was unaware of the situation.*
>
> *A staff nurse is irritated that tasks have not been done in preparation for discharge, Nurse in Charge (NiC) suggests that these have been done. Outgoing NiC chips in with information. Nurses allocated patients who are due for discharge are required to sign off on the discharge. This is a high level of responsibility to check things like swabs have been done. In this case only 2 out of 3 swabs have been done but NiC indicated this was acceptable, staff nurse responsible for discharge was not happy.*

Extended discussion about family not being happy about patient being discharged. Issues flagged for MDT meeting tomorrow morning.

Consultant asking questions. Not happy that a patient has not been discharged yet. The paperwork should have been completed. The discharge nurse is asked to do this (again?).

Answered phone regarding flat cleaning for discharged patient.

Private care packages required – Section 5 care packages – pressure to discharge due to norovirus (outbreak on ward). New package of care requested.

Patient safeguarding issue causing delays with discharge, cannot happen until safeguarding issues are resolved. Family on phone not happy … Updates should be provided by safeguarding team not by nurses on ward. This causes tension.

Interview data provided further insight, indicating that discharge planning was the main focus of all activities on the OPU and involved balancing multiple factors such as reducing length of stay (LOS), ensuring patient safety after discharge, reducing exposure to risks in hospital (infections, pressure sores, lack of mobility), provision of health and social care in the community, the needs and perspectives of relatives and carers, and reducing the risk of re-admissions. This interview excerpt illustrates the challenge of balancing the different goals during the discharge process.

So it's difficult because you set a discharge date, you aim for that, and then if it gets pushed forward it's not that it's not impossible, it just takes up quite a lot of time, because then we have to take on the responsibility of doing things like the section five for social services, or phoning up family members and phoning up care agencies to make sure it's all sorted. Because if you rush discharge, if we haven't checked everything, that we know it's going in, then they'll come straight back in again. So it's the difference between rush, pushing discharge quickly but also making sure it's all set up, which can be what happens when they decide they're going to move a discharge date forward. (Physiotherapist)

In considering the data about discharge, early discussions amongst the research team focused on the need for staff to plan and co-ordinate the different tasks for discharge across different staff groups, agencies, and families and carers. Subsequent observations in the field then focused on how co-ordination was achieved. Later team discussions centred on how co-ordination processes broke down and the consequences of this. The discharge planning narrative contained the following vignette taken from the observational field notes.

An experienced staff nurse, returning to the Nurse in Charge (NiC) role after a holiday, found it difficult to troubleshoot issues associated with the discharge of patients. Time away from the ward meant that the nurse was unable to reconstruct or make sense of decisions that were made. This resulted in time consuming activities such as the reading of all patient notes and making telephone calls. There is a need to maintain knowledge of patient status over weeks in the OPU and time away made it difficult to reconstruct an up to date view of the patient yet this was crucial to ensure timely discharge.

This vignette crystallised our understanding of the problems associated with discharge planning; it was difficult for anyone to have an overview of progress towards discharge and to monitor which tasks had been done, what arrangements had been made, what concerns remained and what tasks were outstanding. Our interpretation was confirmed by clinical members of the team when the narrative was reviewed.

When the narrative was mapped to RE theory, the analysis showed that the RE concepts most relevant to the discharge process were:

- Misalignments between the demand for services post discharge and the availability of those services.
- Variability due to individual patient factors such as the availability of carers, personal preferences, and home environment.
- Goal trade-offs occurred constantly. For example, discharging a patient even if not completely ready may become necessary because of an infection outbreak on the ward, or the stay may be extended because of concerns about safety at home, or discharge may be accelerated so that patients are not kept in at the weekend (when no discharges occur).
- Monitoring the discharge process was difficult because there is no artefact documenting progress that is shared across the different staff groups. There is a white board with a proposed discharge date for each patient but this does not show outstanding tasks or actions and who is responsible for them and it is not always up to date.

Some of these issues can be seen quite clearly in the data extracts provided above but we stress that it was important to undertake a thorough and systematic analysis of the data rather than to proceed on the basis of impressions of the qualitative data. On the basis of this analysis, and discussion with the clinical advisory group, one of our identified quality improvement interventions was to develop an artefact to monitor progress towards discharge that is shared and transparent. The misalignments, variability and goal trade-offs that were identified are not themselves amenable to intervention, but are features of the environment that reinforce the importance of the ability to monitor progress towards discharge. We are currently investigating possible platforms for this, including IT systems and low tech artefacts such as notes and whiteboards.

Discussion

In the previous sections we presented methods for investigating resilience in a clinical setting and provided insights into using the findings to inform quality improvement. We have discussed our ontological perspective and our epistemological position and attempted to illustrate how this informed our data collection, analysis and interpretation. The results show that it is possible to use the results of an exploratory RE study to identify and develop quality improvement initiatives. In ongoing work we will implement and evaluate interventions and it will be important to gather evidence on the effect, if any, on quality outcomes.

A big strength of RE is that it advocates for the importance of understanding the complexities of the whole system rather than focusing on a discrete part (Braithwaite et al., 2013). This is not easy to do in practice and, as researchers, the process has often felt difficult and messy. A strong partnership between clinicians and researchers is one way to help to make sense of the complexities but it is possible that understanding can only ever be partial, shifting and subject to interpretation when studying complex adaptive systems. This should not deter us from applying RE ideas, but it is significantly more challenging than other types of quality improvement work that focus on discrete problems or areas.

Many quality improvement interventions target a particular problem such as falls, and attempt to reduce their incidence. This study, however, was not focused on addressing a defined problem in the OPU, making it very important that we maintained a clear idea of the focus of the project. In hindsight, RE concepts highlighted the need to strengthen processes, which we believe is crucial for proactively preventing problems from occurring in the future. Clinical partners need to understand and support this even though they may be under pressure from managers to address particular problems such as infections or falls. Strengthening process may be a particular advantage of using an RE approach, but the approach could also be applied in a focused way, for example to reduce medication errors. In that case, it would have a narrower focus but the same concepts and steps in the process could be used.

Finally, we acknowledge that the insights this project has generated emerged from an interpretive process and could be criticised as being highly subjective. For example, other researchers may have focused on different issues, highlighted different themes or mapped the data to different RE concepts. However we have been explicit about the reasoning we used at each step, have tested our interpretations with clinicians and within the research team, and have used a systematic well-defined process to interpret our data, all of which reassure us of the robustness of the results. As noted above, insights into complex adaptive systems may only ever be partial and subjective, but the usefulness of the insights and the relevance of the interventions generated can provide evidence of the value of the approach. In due course, evidence of the effect of interventions on quality will also be important. In line with the interpretive epistemological position used, we did not aim to generate general laws about resilience in all clinical settings, although as evidence is accumulated by researchers working in this area it may become apparent that there are some general features of resilience that are common across care settings. This remains to be investigated.

Conclusions

In conclusion, this study has shown that it is possible to use RE insights and theoretical concepts to study a clinical setting and design interventions to improve quality. This is an important step forwards for RE as it has the potential to transform quality improvement thinking if there is guidance for applying it in practice. Future efforts should be devoted to testing this approach in other clinical settings, developing guidance materials, streamlining the process and embedding it in health care as an accepted quality improvement method.

Chapter 5

Using workarounds to examine characteristics of resilience in action

Deborah Debono, Robyn Clay-Williams, Natalie Taylor, David Greenfield, Deborah Black and Jeffrey Braithwaite

Background

Health care organisations (Jordon et al., 2010; Sittig and Singh, 2010) and the profession of nursing (Chaffee and McNeill, 2007) have been characterised as complex adaptive systems (Braithwaite et al., 2013), with uneven and undulating demands and barriers to delivering care. Every day across the world millions of clinicians deliver safe patient care. They do this within complex systems that use traditional approaches to patient safety, approaches that apply linear type solutions to a complex adaptive system, in an attempt to constrain and avoid variability. These solutions often include policies, rules, regulations and technology developed to enforce standardisation of clinical practice. An ability to adapt, predict, adjust and compensate for system barriers and shortcomings are characteristics of experienced and resilient employees (Tucker and Edmondson, 2003; Hollnagel, Braithwaite and Wears, 2013c; Wears, Hollnagel and Braithwaite, 2015a; Braithwaite, Wears and Hollnagel, 2017).

To accommodate for operational failures and deliver care, frontline clinicians use workarounds. Workarounds differ from organisationally prescribed or intended procedures and are employed to circumvent a perceived or actual hindrance to achieve a goal or achieve it more easily (Debono et al., 2013).

It has been argued that, given imposed rigid structures that attempt to model clinicians' work and encode it in rules, workarounds are inevitable (Berg, 1994). Others suggest the possibility for workarounds to provide innovative solutions to problems and highlight opportunities for improvement (Tucker and Edmondson, 2002; Tucker and Edmondson, 2003; Lalley, 2014) and safety (Vincent, 2010). However, the dominant view in the health care literature is that in circumventing safety mechanisms (Patterson et al., 2006; McAlearney et al., 2007; Rayo et al., 2007; Vogelsmeier, Halbesleben and Scott-Cawiezell, 2008), creating unforeseen problems in the system (Mohr and Arora, 2004; Kobayashi et al., 2005) and destabilising attempts to standardise practice, workarounds are detrimental and contribute to error prone organisations (Spear and Schmidhofer, 2005). The prevailing view about workarounds follows a logic that argues: standardisation of practice, enforced by rules, policies, technology and training, supports safety; workarounds undermine standardisation, and can be unsafe. Workarounds have the potential to create an 'underground economy in unsafe practices' (Wachter, 2008: 48).

However, given that most of the time things do go right (Hollnagel, Wears and Braithwaite, 2015; Braithwaite, Wears and Hollnagel, 2015), and given that we know that the use of workarounds is pervasive in health care (Tucker and Edmondson, 2002; Kobayashi et al., 2005; Morath and Turnbull, 2005; Patterson et al., 2006; Koppel et al., 2008; Schoville, 2009; Halbesleben, 2010; Halbesleben et al., 2010; Rack, Dudjak and Wolf, 2012; Debono et al., 2013), then clinicians' workarounds must also contribute to things going right. Workarounds are examples of resilience in action, demonstrating how frontline clinicians continuously adapt to fluctuating conditions and bridge the gap between work-as-imagined (WAI) and work-as-done (WAD) to deliver care (Braithwaite, Wears and Hollnagel, 2017).

Investigating how nurses enact and perceive workarounds, despite threat of potential professional retribution, provides a unique lens for understanding how resilient processes are embedded and managed in the day-to-day workplace, and is the aim of this study. In doing so, it is important to consider local perspective and context. This study examines nurses' workarounds when using electronic medication management systems (EMMS) in everyday practice. The term EMMS encompasses the software (electronic medication administration record (eMAR)) and hardware (desktop computers and computers on wheels (COWs).

A qualitative investigation: the research method

The objective of the study was to explore nurses' enactment and explanation of workarounds with EMMS to conceptualise and understand gaps between WAI and WAD. To understand this issue, a qualitative research method was chosen. Qualitative research methods are used to explore, explain or describe (Marshall and Rossman, 2010) and are valuable in gathering rich accounts of complex phenomena and garnering participants' descriptions and interpretation of events and experiences (Sofaer, 1999).

An ethnographic approach to qualitative research allows for the 'unmasking' of the 'complexities' of clinical work (Hunter, 2008) and in this study used a multi-method triangulated approach to data collection. Ethnography offered the best approach to study workarounds because: actions must be understood in relation to the broader context within which they occur; workarounds are articulation work – mostly hidden from formal accounts of what people do; nurses may be so used to using workarounds that they do not consider them important to mention; workarounds may not be divulged in interview or survey if they are unsanctioned practices because of perceived or real risk of professional retribution. To appreciate work-arounds from the participants' perspective it was imperative to garner nurses' input, ideas and explanations of workarounds observed *in situ*. The researcher's sustained presence in the study settings, and in particular observation of entire night shifts, facilitated participants' understanding of the study, their acceptance of the researcher and open discussion about workarounds. Ethics approval was granted by a Lead Human Research Ethics Committee (HREC) with site-specific ethics approvals gained from the participating hospitals. The University of New South Wales HREC was a ratifying HREC.

Data sources: sampling strategy, study settings, participants

Data were collected in units that used EMMS at two tertiary teaching hospitals in Australia (Hospitals A and B). Different EMMS were used in each hospital. We chose to sample from two hospitals using different EMMS, the number and type of units, models of nursing care, days and shifts to maximise variation (Creswell, 2007) and so minimise the risk of missing major influences on nurses' enactment, explanation and experience of workarounds. The study used purposive sampling, a sampling strategy to select participants able to provide rich information about nurses' use of workarounds with EMMS (Carpenter, 2010). Participants in the study settings were eligible to participate if they were nurses who used EMMS, or were involved in implementation or ongoing support of EMMS (EMMS stakeholders), and rostered on shifts when data were collected. Data were collected between 2011 and 2012; member checking occurred in 2014.

Data collection methods

Investigating WAI: process mapping

Mapping the ideal medication administration process, or WAI, illuminated gaps with WAD and highlighted where workarounds occurred. Process mapping visually represents the steps and decision points in a process (Howard, 2003; Fiore, 2004; Johnson, Farnan et al., 2012). Hospital-specific process maps represented medication administration for a single patient. Their development was informed predominantly by medication administration policy documents and participation in EMMS training sessions. Discussions were also held with EMMS stakeholders about contextual information and interpretation of policy documents. At Hospital A the process map of medication administration was based on route of administration; at Hospital B it was based on whether or not medication administration need to be checked or witnessed. Process mapping also highlighted hospital-specific differences in the interpretation and enactment of medication administration policies. While key to appreciating aspects of

WAI, the process map represented the linear process for medication administration to a single patient and did not reflect the complexity of WAD; therefore additional data collection methods were required. For example, nurses often prepared and checked medications for more than one patient at the same time before moving on to administer them.

Investigating WAD: a triangulated approach

The study used a triangulated approach (O'Donoghue and Punch, 2003) to WAD data collection including observation (non-shadow and shadow), interviews and focus groups. Using several data collection methods (triangulation) provided different perspectives of nurses' enactment and explanation of workarounds.

Observation

This study used observation to capture actual (WAD), rather than prescribed (WAI), behaviour (Tucker, Edmondson and Spear, 2002; Tucker and Edmondson, 2003). Observation has been demonstrated to be a valid and reliable data collection method to apprehend what is occurring in a particular setting (Dean and Barber, 2001). Observation was used to explore whether or not nurses used workarounds and the context within which they did so, or chose not to do so. A comprehensive note-taking strategy used to collect data was guided by Spradley's (1980) list of generalised 'concerns' including: the physical space; the people involved; the physical objects; the single activities that people performed; the sets of related activities that people performed; the sequencing of activities over time; the goals of the activities; and the expressed feelings (Wolfinger, 2002).

The types of skills and professional characteristics that were important to nurses in this study became apparent through what they individually and collectively praised and criticised. Observation was essential to capture this information because communication of the information and agreement or disagreement with a shared view was frequently non-verbal, including rolled eyes, raised eyebrows, sarcasm and inclusion or exclusion in activities. It was also useful to capture this information over a period of time to see whether there were patterns – was it just one nurse who propagated a particular view or attitude, or was it shared by others and collectively perpetuated? Did it change according to context? Did it differ between units?

To capture resilience in action it was important, as much as possible, to observe nurses across entire shifts, rather than just the medication administration process. Workarounds often led to, or were caused by, other workarounds either in the same or a separate process. For example, nurses co-checked medications for multiple patients early in the shift when they anticipated they would get busy later (primary workaround). They then wrote the patient identification information on a piece of paper and used this as another check later in the shift when administering the medication (secondary workaround). Cycles of nursing work (Tucker et al., 2002) mean that what happens in one part of the shift may influence what occurs in another part of the shift. It was also important to observe different shifts, including after hours, to better understand the complexity of the system – what happens in one part of the system at one point in time will influence what happens in other parts of the system.

An important part of the observational component included participants' *in situ* explanations for observed contextual features, activities and interactions, which contributed to a description of the characteristics of each setting, events and reasons for behaviour from their perspective (Fetterman, 1998). This was essential because, as with the difference between WAI and WAD, attributing a motivation to observed action may be WAI. In the same way as policy makers imagine how work is done, researchers need to be careful not to impute reason as imagined to observed behaviours. For example, one nurse explained that whether he/she used workarounds depended on with whom he/she was working. Unlike other participants, the nurse's explanation was that it depended on his/her perceptions of how competent his/her colleague was rather than how official or strict they were. Offered rationalisations should not be assumed to be generalisable to all nurses in every context. Rather, it is important to identify how participants conceptualise or rationalise their behaviours in a given context rather than imagine what it is.

Non-shadow observation

The process of ethnographic observation includes initial relatively unfocused observations, moving to more focused observation as the researcher becomes more familiar with the setting (Spradley, 1980). Non-shadowing observations of nurses were used to build a picture of the normative behaviours, attitudes, interactions and beliefs in the study settings. The researcher observed: corridor and nurses' station activity; communication patterns; handling of artefacts (the medication system, the notes and equipment), rituals, symbolic behaviour and interactions. Participants explained observed activities, conversations and interactions. During this period of data collection, the new and novel quickly became familiar and less noteworthy to the observer. It was therefore important that during initial non-shadow observations on each unit, the researcher took detailed notes.

Shadowing observation

During shadowing observations nurses were closely followed as they conducted medication administration, medication-related work and other activities including interactions with computers, ordering and replacing stock, discussions with other staff and handover. Shadowing observation was used to identify the occurrence or non-occurrence of workarounds in the medication administration process, contextual factors in which workarounds were embedded and the ways in which nurses managed their own, and responded to their colleagues' use of workarounds. Participants were asked about observed variations in the medication administration process, to identify whether the variation matched the definition of a workaround used in this study. Nurses who were not being shadowed also offered perceptions about their experiences of using EMMS.

Individual and focus group interviews

The purpose of interviews and focus groups was to explore the reasons and meanings participants attributed to their own and their colleagues' medication administration behaviours. Interviews investigated individuals' explanations and experiences of their own and their colleagues' workarounds. Shared explanations and experiences of the use of workarounds were exposed, using group interaction to generate ideas during focus group discussion (Kitzinger, 1996; Morgan, 1996). Interviews with 44 participants were transcribed from digital recordings; two were transcribed in real time at the time of the interview. All focus group data (N=7) were transcribed from digital recordings.

Data analysis

A general inductive approach to qualitative data analysis was used in this study (Thomas, 2006). This approach used inductive analysis to develop key themes and processes that emerged from the data, and deductive analysis, which framed the analysis against the research questions. The interview data were coded using QSR NVivo 10.0 Software. The annotation feature of NVivo 10.0 was useful during coding to record insights and reflections. After reading through the data and making notes on patterns, thoughts and ideas, the interview data were examined for behaviours that matched the definition of workarounds. The observed or described behaviour (WAD) was compared with the ideal process (WAI). Coding the data for workarounds highlighted the importance of identifying the goal of the workaround. The process then involved identifying the challenge (or challenges) to meeting that goal or meeting it easily. Often there were multiple goals and the salience of achieving one goal was foregrounded over others depending on the context. There were also factors that moderated whether or not workarounds were used to achieve particular goals. In order to determine whether a workaround had manifested, it was necessary to ascertain the goal of the behaviour and the perceived workflow hindrance. A goal might be the outcome of the entire process, or the outcome of each process step. The data were then examined for themes illuminating how nurses rationalise, perceive and experience identified workarounds. After coding the interview data for themes and categories, entered into a Microsoft Excel file, observational field notes and focus group data were interrogated for categories that had been identified in the interview data, while staying open for the emergence of new categories.

Interpretation

The study was not designed to examine nurses' use of workarounds by applying predefined theoretical constructs. An explanatory framework of nurses' use of workarounds with EMMS was generated based on concepts that emerged from the study findings. Support for interpretation (see findings) was offered during member checking sessions at the participating hospitals.

Validity and verification

According to Senge et al.'s (1994) ladder of inference, we selectively attend to information, and our mental models influence the way we interpret it and the conclusions we draw. When conducting research it is imperative to use strategies to recognise and manage the influence of those mental models. Validity strategies included reflexivity, peer debriefing, searching for disconfirming evidence, and triangulation (Creswell and Miller, 2000). Strategies used to establish credibility of the study findings included clear exposition of data collection and analysis methods, audit trail and peer review; and prolonged engagement, persistent observation and member checking (Creswell and Miller, 2000).

Findings

Five consistent categories were identified in relation to nurses' use of workarounds. For the most part, they used workarounds to be or to be perceived to be: 1) time efficient; 2) patient-centred; 3) a team player; and to 4) maintain patient safety individually and collectively. However, on occasion, nurses also used workarounds because of 5) contextual limitations. Spanning all of those was knowledge and experience, particularly in relation to knowing when it was acceptable to use workarounds and when it was not safe to do so. At times, workarounds to achieve one goal (primary workarounds) undermined achievement of another goal and so led to secondary workarounds.

The same workaround may have been used to achieve different goals depending on the context. That is, while the workaround was unchanged, the barriers to and the goal of the workaround differed. We provide an illustrative example below.

Policy required that nurses take an active eMAR on a COW to the patient when administering medication. In some instances it was not possible to take a working eMAR to the patient because of contextual limitations (e.g., there was not wireless connectivity). At other times nurses did not take the COW to the patient (primary workaround) to circumvent barriers to achieving different goals (to be time efficient, patient-centred, a team player, safe). Nurses used secondary workarounds to verify patient identity when they did not take the COW to the bedside. These included: writing identity information on a piece of paper; memorising the patient medical-record number; relying on familiarity with the patient and their medications; using the bed number, or the name above the bed, to cross check patient identity; or working from print outs of medication orders.

Time efficient

Nurses worked around the barrier to being time efficient caused by having to manoeuvre a COW to the bedside. Instead they checked off the medication in the eMAR on the COW parked in the corridor or at a desktop computer. The importance of time and of being time efficient was reinforced on many levels. There were set medication administration times, meal times and shift times. Nurses derided colleagues who were slow and applauded those who completed their tasks on time. Nurses were perpetually preparing for unexpected events. The EMMS increased the visibility of medication work and some nurses described an emotional reaction to a visible overdue medication alert (OMA) and workarounds they used to remove it because it was perceived to signal that they were late and had 'failed'. There was noted variation in the response to the OMA between units and hospitals.

> *I have noticed that if people do it differently if they are busy or not busy.* **When they are busy** *both of them might not go, or* **the computer won't go into the room**, *or they won't wait until the tablet has been taken before they start writing, sometimes they might forget to double check the MRN – which they are usually pretty good with with S8s and S4s but not usually regular meds. (Focus Group 3)*

Patient-centred

To be patient-centred the COW was not taken to the bedside, particularly in a four-bedded room, at night because it was noisy and the screen was bright and woke all of the patients in the room. There were guidelines (NSW Health, 2009) that reinforced the importance of patient-centred practice – of keeping noise to a minimum at night for the patients' benefit, and nurses were observed to whisper, walk softly and work quietly.

> *Sometimes you leave it then too because there's no real point grabbing your clunky machine waking everyone up, as you're dragging it down the hallway, to park it, to have the bright light shining and you confuse patients, that's why they wake up. (Interview: Nurse_39)*

Teamwork

Delivering patient care was observed to be heavily reliant on teamwork. To deliver safe, patient-centred care in a timely manner, nurses needed to work as a team, to ask for and offer help to colleagues. Being a team player included helping and not questioning senior colleagues when they were busy, looking after allocated patients and not impinging on teammates' time. Ideally WAI nurses would be able to maintain continuity with the COW they were using for the medication round. However, during medication rounds doctors were described and observed to 'hijack' COWs. Participants explained that when there were not enough COWs, rather than take one that someone else was using, or to work around a potential argument with the doctors about a COW, they flexibly changed their behaviour and signed off medications at the desktop computers.

> *I've got the computer at the desk and written down the next patient's medications and the dosage and then gone into the room and given those . . .. Because I don't know how long it's going to be before I'm getting my computer back, yeah. (Focus Group 8)*

Patient safety

Nurses did not take a COW to the patient bedside when they judged that to do so compromised patient safety. According to infection control policies, COW should not enter isolation rooms or should be cleaned with detergent and disinfected prior to leaving the room. To maintain patient safety individually and collectively, when a COW could not be dedicated to an isolation room, nurses worked around the potential risk of cross infection by not taking COWs to the bedside of an isolated patient. Similarly, in other situations to maintain safety, a COW was not taken to the bedside if to do so created a falls risk for elderly patients.

> *The trolley. If no – if they find out it's too much equipment, too many furnishings in the room and it's high risk for a fall for the patients, they can leave it outside and get the drawer. Just take the single drawer, put it*

on the COW and dispense the medication, put it back, check their MRN number and go to the patient and give it. (Interview: Nurse_42)

Contextual limitations

In some instances nurses worked around the lack of wireless connectivity (black spots) at a bedside by leaving the COW at the doorway, where it remained connected. When there were not enough functional COWs, nurses were observed to use desktop computers.

Nurse 86 pushes the cow next to bed 14. The laptop loses connectivity and the electronic medication system freezes again in room 14. Nurse 86 pushes the laptop out into the corridor and goes to the four bedded room across the corridor intending to give their medications next. However he cannot log back in and so he goes and finds another computer - he says 'now I'm getting irritated'. (Observation 86)

Discussion

This study examined nurses' enactment and explanation of workarounds to conceptualise and understand the gaps between WAI and WAD. Nurses' work was observed to be a continual juggle use of the EMMS with: critical decision making; multi-tasking; interruptions; operational failures; relationships; time pressure; managing expectations; and teaching activities within a context shaped by ward and professional culture. Medication administration forms one component of nurses' work. This chapter illuminates attributes of resilience in action. It demonstrates how nurses adapt, flex and navigate competing demands so as to adjust under expected or unexpected conditions in order to sustain required operations. It highlights the shifting and jostling demands of delivering care that prioritise one goal over another in a continually changing way, the role of context in influencing that process, and ongoing judgements about when to use [or not use] primary and secondary workarounds. In an ideal WAI world, nurses would be able to simultaneously realise all of these goals at once. Structures and processes would be in place that would support nurses being simultaneously time efficient, safe, patient-centred, a team player, and there would be no contextual limitations – they would not have to choose or juggle. In reality, nursing and other health professional work is complex and there is a need to continuously navigate competing demands and make judgements about what is most important in that moment. At times, workarounds to achieve one goal (primary workarounds) undermined achievement of another goal and so led to secondary workarounds.

Nurses described a disconnect between policies created for use in the ideal world (sacred (Durkheim, 2001) (WAI)) and the feasibility of their use in reality (profane (Durkheim, 2001) (WAD)) of delivering care in a complex adaptive system. The tension between top-down pressures from the external environment and the bottom-up pressures of operationalising everyday work contribute to the persistence of workarounds and nurses used workarounds to reach a balance between the bottom-up challenges of day-to-day work and 'top-down' pressures, including policy directives (Azad and King, 2012).

WAI and WAD also apply to the research process: research-as-imagined versus as research-as-done – how researchers perceive research will be done and how it actually happens in the field. In every system, differences between WAI and WAD need to be considered. In the practicalities of delivering health care or in researching the provision of health care – there will always be these differences and allowances made, or workarounds developed, to manage the gaps between the two.

It is imperative to consider multiple layers when investigating and interpreting the clinicians' explanations for their behaviours. Nurses, for example, may offer time pressure as a reason for not taking the COWs to the bedside, yet may not appear to be busy. Upon investigation, time pressure can be understood to be immediate and anticipated and the importance of completing tasks and having time to be ready to assist colleagues or cope with unexpected events central to what it means to be a good nurse. Sitting at the nurses' station having completed all of one's tasks, for example, may be seen in some units as evidence of a nurse's efficiency. Unless these aspects are also taken in to consideration, conclusions will represent an artificial model that presents WAI.

Implications and conclusions

When introducing technologies, effort is frequently expended to consider the potential impact of new technology on workflow. WAI is based on the idea that if there are enough resources, policies to dictate use, and education then there will be no reason why nurses would not use EMMS as imagined. However, the potential impact of introducing new technologies and policies on professional image, the construction of which is mediated by society, individual and collective beliefs and expectations and experiences (Bourdieu, 1984), is given less attention. Social and professional culture and influences should be considered as part of context. Workflow problems are interpreted differently depending on: what is considered to be most salient in that moment; professional experience; who I am working with; and beliefs about what constitutes being good at one's job. Workarounds enabled nurses to achieve the characteristics that are important to nurses doing a good job. Changes such as the introduction of technology and policy that challenge or illuminate nurses work (e.g., EMMS makes work more visible) will have an impact and may lead to workarounds.

Nurses' workarounds optimise practices, processes and systems that can never fully be prescribed even though large organisations and governments seek to do so. As examples of first order problem solving (Tucker and

Edmondson, 2003) workarounds often address immediate but not underlying problems and the system appears to be in balance. However, over time, repeatedly working around problems can lead to frustration, fatigue, burnout and reduced capacity to address underlying problems (Tucker and Edmondson, 2003). Therefore, while on the one hand workarounds are localised acts of resilience, in as much as they prevent organisational learning by covering problems (Tucker et al., 2002; Tucker and Edmondson, 2003), and reduce the ability of an organisation to gauge what is happening within itself, on the other hand they may contribute to brittleness of the system as a whole. Our work supports Holden et al. (2013) and McLeod, Barber and Franklin's (2015) suggestion that identifying nurses' workarounds can be used to locate potentially suboptimal processes and systems (Holden et al., 2013). Because they are effective in solving problems, however, they can also hide problems. Therefore to use workarounds to identify suboptimal processes, they must be examined intentionally. Harnessing localised acts of resilience by making them visible and building them into collective organisational awareness and use will contribute to organisational resilience. We need to consider how to explicate components of organisational resilience such that organisations can build on localised acts of resilience. We need to better understand when workarounds are deemed to be acts of localised resilience and when they are not – including who can do them, when and where.

Chapter 6

Simulation as a tool to study systems and enhance resilience

Ellen Deutsch, Terry Fairbanks and Mary Patterson

Background

Simulation, which includes replicating portions of patient care in a realistic and engaging manner, provides a powerful tool to explore the characteristics of resilient health care (RHC), and to bring RHC into practice. Health care is innately dangerous, balancing the risks of medical conditions with the risks of treatment. Managing this risk successfully requires knowledge, skills, resources and flexibility. Because of the complexity of health care delivery, both requirements and resources are dynamic, and resilience is desirable (Braithwaite, Wears and Hollnagel, 2015; Dietz, 2009; Hollnagel, Wears and Braithwaite, 2015). Resilience has been defined as the 'ability of the health care system (a clinic, a ward, a hospital, a county) to adjust its functioning prior to, during, or following events (changes, disturbances, and opportunities), and thereby sustain required operations under both expected and unexpected conditions' (Wears, Hollnagel and Braithwaite, 2015a).

This chapter will address a specific application of simulation to support RHC, illustrated by the example provided in the case study as well as the variety of ways that simulation can be used to help understand and support the emergence of RHC. Health care simulation replicates aspects of patient care conditions so as to provide opportunities for health care providers, or organisations, to learn (or assess), at the participants' relative convenience, without direct risk to patients (Deutsch, 2011; Gaba, 2007; Kneebone, 2003). New applications of health care simulation have been inspired by improved simulation technology, better understanding of adult learning processes and, recently, increased appreciation for the relevance of human factors and systems engineering principles applied to health care delivery. Simulation is used to improve the knowledge, technical skills and non-technical skills of individual health care providers; and to improve the knowledge, technical skills and non-technical skills of health care teams.

The next frontier for taking advantage of the diverse potential benefits of simulation involves the use of simulation to assess and improve the socio-technical systems that impact health care delivery; this aligns nicely with the exploration of concepts related to the emergence of organisational or system resilience. The quantity and relevance of information derived from simulation experiences would be difficult to obtain by more conventional means including traditional event debriefings, focus groups, 'walkthroughs', scavenger hunts, surveys or by analysis of large data sets. Simulations can be repeated as desired, to learn how to function within the local health care delivery system, as well as to support iterative improvements in patient care processes. Each simulation's theoretical construct – its 'story' – provides the clinical context for the simulation. Participants willingly suspend disbelief and become engaged in the activities of the simulation.

Simulations can be conducted in simulation centres or *in situ*. Simulations conducted *in situ* approach an accurate representation of 'work as done'. Simulations with real teams functioning in real locations allow intentional probes of actual patient care processes and the enveloping socio-technical system (i.e., at the macro, meso or micro level). Even if the simulation is not intended as a system probe, relevant findings may be serendipitously recognised. For some threats to patient outcomes, knowing the correct medical response is essential but not sufficient. Knowing how and being able to activate responses and obtain resources in the actual clinical setting is also necessary.

With the exception of simulations designed for summative testing, feedback or debriefing is an important component of simulations. Several debriefing styles and philosophies are practised; their commonality is generally an attempt to elicit observations and insights developed by the participants, with guidance from subject matter experts (Eppich and Cheng, 2015). Debriefings may include video review of actions that occurred during the simulation. Some of the important lessons provided by simulation are affective, based on the tone set and modelled by the facilitator, such as respectful curiosity, and valuing every participant's contribution. Learning from a simulation is not limited to direct participants, but is also available to the simulation's facilitators, observers, reviewers and others.

Simulation to support resilience: opening a new health care facility

The post-tonsillectomy haemorrhage case is one example from a series of simulations conducted to prepare to open a new satellite, low acuity, paediatric community hospital approximately 40 km from the organisation's quaternary care main paediatric hospital. In addition to building a new facility, a significant number of new staff, some without paediatric experience, were hired for the satellite hospital; and care teams and processes were planned in new configurations. Though not designated as a trauma centre, the satellite hospital is located at the junction of two major highways.

Leading up to the opening of the satellite hospital, the simulation team was asked to evaluate both the facility and health care provider teams for clinical readiness, and to identify and help mitigate Latent Safety Threats (LSTs) and system issues in the new facility. The investigators identified high-risk critical scenarios that were, in their opinion, likely to occur based on the location and branding of the satellite hospital as a 'Children's Hospital', including trauma scenarios and scenarios requiring simultaneous resuscitations of multiple patients.

The design of this project includes clear examples of the four main activities of resilience engineering: monitoring, reacting, learning and anticipating (Hollnagel, 2011a). The methodology and analysis incorporated a Safety-II approach of reinforcing appropriate actions and resources (Hollnagel et al., 2015; Woods and Cook, 2006), making the margins and constraints of the system visible, and developing team behaviours that have the potential to improve the adaptive capacity of the team (Braithwaite et al., 2015). The concepts of margins, constraints and boundaries are based on the work of Cook and Rasmussen (Rasmussen, 1997; Cook and Rasmussen, 2005), which suggests dynamic trade-offs between pressures to optimise workload, productivity, and the boundary of safe performance.

Case study

A 3-year-old male presented to the satellite Emergency Department (ED) with bleeding from his right tonsillar fossae 4 days after undergoing uneventful adenotonsillectomy (T&A). The patient was triaged and placed in a resuscitation room, and the Otolaryngology and tertiary centre Emergency Care physicians were alerted. In addition, the on-site transport team was activated. A history was obtained from the child's parents, a physical examination was completed, an intravenous (IV) line inserted and a 'type and cross' blood specimen obtained to prepare for possible transfusion. As there was not an ENT specialist available at this hospital, the transport team was quickly dispatched to take the patient to the tertiary care hospital 40 km away.

Debriefing after the event included ED physicians, nurses, paramedics and organisational leadership. LSTs were identified, such as unavailability of defibrillators. Participants discussed teamwork, workload, access to resources and other factors that might impact patient care. Facility administrators participated in the discussion as the group identified possible system weaknesses and potential solutions.

The 3-year-old, a high-technology manikin, and his 'parents', scripted simulation 'actors', returned to the hospital entrance 3 weeks later, and the scenario was repeated.

In all, simulations addressing 24 critical scenarios were conducted prospectively, prior to opening a new emergency department. Simulation sessions were designed in two phases: two sessions occurred 10 days apart at a simulation centre, followed by two sessions 3 weeks apart *in situ* (in actual patient care locations), in the actual satellite emergency department. The sessions focused on identification of LSTs as well as developing process and system information. Because the project was structured to allow evaluation of unintended consequences, the second round of *in situ* simulation included a deliberate focus on the consequences of system modifications that had been implemented following the initial round of simulations. Administrators and clinical personnel were able to mitigate 32 of 37 (86 per cent) latent safety conditions identified during *in situ* simulations before opening the new ambulatory surgical care facility for actual patient care (Geis et al., 2011).

Study methods

The study was conducted in two parts. Over a 3-month period, four centre-based simulations in each of two sessions (eight simulations total) were used 'to define roles and scope of practice', followed by eight *in situ* simulations in each of two sessions (16 simulations total) focused on 'evaluating the quality of initial solutions and identifying any unintended consequences' of applying the developed solutions (Geis et al., 2011). Immediately following the simulation, participants completed the National Aeronautics and Space Administration-Task Load Index (NASA-TLX; Hart and Staveland, 1988; Weinger, Reddy and Slagle, 2004; Becker et al., 1995; Gregg, 1994), to measure perceived workload. Then a group debriefing was guided by an experienced facilitator, incorporating video review of participant actions, and input from a human factors expert. A post-debriefing survey sent electronically to all participants offered an opportunity to share information that participants may have been uncomfortable sharing with the group, or may have thought of following the debriefing. Finally, team behaviours were scored by trained video reviewers using the Mayo High Performance Team Score (Malec et al., 2007). Clinicians who were not participating in a particular *in situ* simulation observed the simulation and participated in the debriefing in a 'fishbowl' methodology, which allowed observers to contribute their additional insights about the work and enriched the debriefing.

Additional details of the methods and results are provided in the article by Geis et al. (2011); the intent of this chapter is to provide a brief summary of how the methods and results can be interpreted in the context of concepts related to resilience.

Study data sources and rationale

Information gathered from the simulations of a patient with a post-tonsillectomy haemorrhage, and other patient care simulations, was evaluated using a mixed methods approach, which included both qualitative and quantitative analysis to triangulate conclusions and provide reports to leadership. Based on previous simulation experiences, the investigators had a strong bias that some of the most important information uncovered occurs during the video analysis and guided participant debriefing. The information and insights derived during guided debriefings immediately following a simulation scenario would be difficult to obtain in any other manner. However, it was also recognised that organisational leadership is influenced by 'hard data'. Therefore, a mixed methods approach was chosen to ensure that the richness of the qualitative data as well as the perceived strength of numerical data was included.

During debriefing, frontline staff provide detailed information and insights concerning what aspects of health care delivery worked and what did not work during the simulations, and frequently offered possible solutions. Unique teams participated in the same scenario, allowing identification of common insights and themes.

The NASA-TLX (Hart and Staveland, 1988; Weinger et al., 2004; Becker et al., 1995; Gregg, 1994), a validated scale recommended by the study's human factors expert, was used to assess the boundary conditions and the limits of safe performance, in different settings and with different team configurations. This provided an indicator to the simulation team and unit leadership about when workload was high, and margin – or adaptive capacity – was diminished; as well as information about which processes and systems resulted in more optimal (moderate) workloads.

At the time this project was conducted, there were very few team behaviour scales that were intended for inter-professional use and were not focused on a single discipline. The MHPTS was the only validated scale at the time that was intended to evaluate the performance of the entire team during an acute event. Other tools, such as Anaesthetists' Non-Technical Skills for anaesthesiologists, were intended to evaluate non-technical skills but not the team behaviours of other team members (Fletcher et al., 2003). For this reason, the MHPTS was used to score team crisis resource management (CRM) behaviours during video reviews. The use of the MHPTS allowed comparison of team performance under different conditions and team configurations, including team management of unexpected challenges (e.g., the simultaneous presentation of more than one critically ill patient).

Data sources and data collection

As mentioned previously, this was a mixed methods investigation. All sources of data were obtained from the 24 interdisciplinary simulations that were conducted as part of the project. The NASA-TLX scores were completed on paper by participants immediately following each simulation and require no longer than a couple of minutes. The largest sources of data were the guided debriefings that occurred following each simulation. The facilitators captured this information on paper and video recording. These were the source of much of the information concerning the work and the capacity of the team.

In addition, an email was sent to all participants following the simulation in case they wished to share information that they had not remembered or did not feel comfortable sharing during the debriefing.

Video recordings of the simulations were analysed by trained reviewers using the MHPTS to evaluate the team performance of these newly formed teams both in the centre-based, and *in situ*, environment.

A unique aspect of this project was the planned re-evaluation of the systems following changes made as a result of the simulation results. Changes in team configuration, resources and facility layout were evaluated using simulation to assess any unanticipated consequences.

It should be noted that this approach required a significant investment in time and resources to accomplish. The centre-based sessions required a total of 8 hours – two 4-hour sessions separated by 10 days to allow leadership to make modifications to the team composition. Two additional 8-hour *in situ* simulation days were separated by 3 weeks, again to allow for modification of the processes and environment. Each simulation required a team of simulation specialists (four to five), a human factors consultant and the clinicians involved in the simulation. Typically four to six clinicians might participate in a simulation. Additional time was invested by the facilitators in analysis of the qualitative data and by the video reviewers. During the *in situ* days, three to four teams of clinicians were present. This represents an impressive investment of resources for this project, which was funded by the organisation. The investigators believe that by the time this project was developed, the simulation programme had already demonstrated significant value to the organisation and the organisation was therefore willing to invest in this because this type of facility was an entirely new endeavour for the organisation and in the hope of avoiding significant safety issues when the facility opened for patients.

Summary of study results and data analysis

The data obtained fall into several categories. First, from the analysis of the information obtained during the debriefing, we were able to identify several themes concerning team performance as well as specific LSTs. During the centre-based simulations several teams identified that the medical team leader needed to act solely as team leader and not become involved in procedures if possible. This led to a reconfiguration of team responsibilities such that a nurse, rather than a physician, became responsible for placing an intraosseous (IO) line. The centre-based simulations also revealed knowledge gaps among new personnel related to paediatric care as well as gaps in team behaviour expectations relative to the organisation. These included fundamental gaps related to weight-based dosing and paediatric resuscitation, as well as lack of familiarity with the concept of a shared mental model or stepback.

During the simulation centre sessions, the NASA-TLX scores revealed a trend suggesting that the medication nurse had the highest workload. This became obvious during the *in situ* simulations, in which the medication nurses consistently scored >60, in the high workload potentially unsafe range (Gregg, 1994; Weinger, Reddy and Slagle, 2004). This was a surprise to leadership and demonstrated the adverse impact of the physical layout as well as the need for a backup nurse to assist with multiple complex weight-based medication calculations. Data from the NASA-TLX scores, video-recording analysis and the debriefing analysis convinced organisational leadership of operational hazards and resulted in a change in the physical configuration of the medication nurse work areas as well as augmenting role assignments with a second nurse or a pharmacist to assist with obtaining medications emergently.

One of the most interesting LSTs identified was the absence of sufficient oxygen pressure to ventilate two patients simultaneously in the resuscitation bay. Though not designated as a trauma centre, the investigators believed it was highly likely that there would be times that two critical patients would be present simultaneously. Therefore this was one of the scenarios tested *in situ*. It was surprising to the entire team that the oxygen pressure was inadequate to support two patients simultaneously. This threat was remedied prior to opening the facility. It is unlikely that this type of threat would have been identified without simulation. In fact, subsequent to the facility opening, the resuscitation bay had been used by two critical patients on many occasions, and without understanding and improving system capabilities through the use of simulation, it is conceivable that patients may have been harmed in this situation.

Altogether, 37 identified LSTs were identified, and categorised broadly as: equipment, medication, personnel and resource issues (Geis et al., 2011). Additional findings with particular relevance for resilience include:

- Conditions that contributed to cognitive fixation and loss of situation awareness (Wright, Taekman and Endsley, 2004; Endsley and Kaber, 1999).
- Worse teamwork during *in situ* (versus simulation centre) simulations, suggesting that optimising teamwork is more challenging in an *in situ* simulation.
- Consistently high workloads for nurses responsible for providing medications, particularly during scenarios involving multiple patients.

Interpretation and discussion

This project demonstrates the value of simulation to help understand s work as done and identify potential systems and resource issues that may enhance or limit team and system performance. Proactive simulation supports the emergence of system information that would not be obtained in other ways, short of a clinical disaster.

Analysis of this simulation project allows exploration of the four abilities of resilience: to monitor, respond, learn and anticipate (Dekker et al., 2008).

- Monitor: participants learned and practised observing the patients' conditions and the actions and capabilities of other team members. Participants observed that the physician team leader was fixated on procedural tasks, and responded by implementing changes in team members' responsibilities. This included expanding the scope of the nursing role to permit nurses to place IO needles for vascular access.
- Respond: participants and leadership adjusted the resuscitation team structure and responsibilities based on information learned during the simulations. In addition, the organisation addressed the 37 identified LSTs urgently prior to the facility opening. In the case of the inadequate oxygen pressure described previously, a second oxygen line was installed in the resuscitation bay prior to opening the facility.
- Learn: participants practised technical skills, such as cardiopulmonary resuscitation, and IO line insertion; and non-technical skills; and the organisation has developed educational curricula for specific skills. In addition, participants practised how to function together in a coordinated, effective manner. Despite this training the MHPTS scores decreased from the centre-based setting to the *in situ* setting, possibly reflecting higher levels of individual uncertainty and stress in the actual clinical setting. The organisation recognised that a 'one and done' approach to teamwork was not likely to be successful. Therefore, the organisation requires 4 hours of simulation based team training for new employees, and inter-professional *in situ* simulation training is conducted for all clinicians on an ongoing basis at this facility.
- Anticipate: practising the management of a variety of critical emergent patient conditions improves participants' understanding of how these conditions evolve, what treatment may be required, and how patients may respond to interventions. Perhaps more importantly, the investigative team worked to anticipate the types of emergent clinical scenarios that were likely to occur in this setting, despite the official designation of this facility as a community health care facility. This led the investigators to develop challenging simulations that would test the adaptive capacity of the system for these unsolicited but possible events. It would be difficult to obtain the quantity and relevance of information derived during the simulation, including the debriefing, by more traditional means of evaluating complex systems and events, such

as by using interviews, surveys, focus groups or other processes that tend to reveal 'work as imagined'.

Information obtained from these simulations convinced unit leadership that the adaptive capacity of the previously planned team configuration was limited and likely to lead to an error. For example, information from this project resulted in the reconfiguration of the resuscitation team in order to provide additional resources to the medication nurse. More generally, the workspace and the resuscitation team were reconfigured to optimise performance and improve the potential for resilience in critical situations.

Simulation conducted *in situ* provides a powerful mechanism to approach understanding and anticipating work as actually done, in order to better understand boundaries and margins. Although the new facility that was the subject of the study had been expertly and attentively planned, the complexity of health care precludes a complete understanding of all of the socio-technical factors impacting patient care. For example, the simulation of resuscitating two patients at the same time revealed that the oxygen pressure in the resuscitation bay was inadequate to provide assisted ventilation to more than one patient simultaneously. This is a critical example of identifying system limitations and improving margin by modifying the system. Of interest, without the conduct of this type of simulation prior to clinical occupancy, it is unlikely this boundary would have been recognised before a patient was exposed to a significant risk.

A separate study based on *in situ* simulation was prompted after a (real) patient presented to a paediatric emergency department with partial airway obstruction. In this case, the index patient survived without harm (a near miss), but rather than the relief associated with 'good luck', the team used this event as impetus to evaluate the 'difficult airway' response system, which is critical but infrequently activated. Using simulation to replicate the emergency, the equipment, personnel and protocols were studied and modified. The result was that the difficult airway team was reinvented with attention to socio-technical considerations, and in a manner that recognised the physiologic boundaries of the paediatric patient with a difficult airway crisis (Johnson, Geis et al., 2012).

Simulation: the medium is the message

The studies described above illustrate concrete applications and concrete results of health care simulation implemented in a manner that supports the emergence of resilience, and described in terms that reference resilience theories. Discussion has addressed 'why' and 'how' specific tools (e.g., NASA-TLX) can be used during simulation to investigate hazards and evaluate potential responses. The information obtained was qualitative as well as quantitative, and unanticipated findings were not just accepted, but sought. Nonetheless, the contributions of simulations were analysed in the context of directed goals.

Regardless of, or even without, directed goals, in many simulations the process of participating provides affective lessons that align with the four stated abilities of resilience – the ability to monitor, respond, learn and anticipate (Wears, Hollnagel and Braithwaite, 2015a). Just as resilience emerges from a combination of conditions; the four abilities of resilience often emerge from simulation. This occurs regardless of the scale and structure of the simulation, which can range from a simple physical model used by one person to address a discrete skill, to a high-technology, interactive human model situated in a clinical scenario involving a group of people; or numerous other variations.

The ability to monitor, respond, learn and anticipate is espoused in the content of simulations; these abilities are practised during simulations, these abilities may be discretely articulated during simulations, and these abilities are demonstrated during simulations, particularly during the debriefings. Similar to a concept stated by Marshall McLuhan, 'the medium is the message' (McLuhan, 1964). It would be difficult to constructively engage in a simulation and not improve the ability to monitor, respond, learn or anticipate, for an individual, or for a system.

Conclusion

Simulation offers an opportunity to replicate medical conditions – be they routine, or rare or dangerous – in a realistic and engaging manner, which supports the emergence of resilience. Though simulation has been proven to be effective as a tool for improving the capabilities of individuals, teams and health care delivery systems, the outcomes from this work are not often described or interpreted in a resilient health care framework that explicitly identifies resilient characteristics, or system boundaries, margin and adaptive capacity. As demonstrated in the examples provided in this chapter, simulation can support and enhance the ability to evaluate and understand health care delivery in new, renovated and existing environments to a degree and with results that are not possible with other techniques.

In safety science, there is a need for deliberate attention to the techniques that support our ability to understand, test, develop and improve the adaptive capacity of individuals and teams, and the systems they function within. As we explore real-life opportunities to understand and describe clinical work in a resilient health care frame, we should take advantage of simulation's unique abilities to help us study and enhance resilience.

Chapter 7

Exploring resilience strategies in anaesthetists' work

A case study using interviews and the Resilience Markers Framework (RMF)

Dominic Furniss, Mark Robinson and Anna Cox

Introduction

Resilience strategies refer to a set of behaviours that lie largely outside of the formal system (e.g., outside of standard operating procedures) that are invented, adopted and adapted to maintain safety and performance (Furniss et al., 2011b). Understanding resilience strategies brings us closer to understanding the informal behaviours and practices that contribute to the safety and performance of systems before they go wrong. However, there are few methods to uncover such behaviours and practices. In this chapter we explore the potential for the Resilience Markers Framework (Furniss et al., 2011b) to provide guidance for development of an interview script aimed at uncovering resilience strategies used by anaesthetists. We were particularly interested to investigate whether resilience strategies, as a concept, is understandable and relevant to anaesthetists, whether anaesthetists exhibit resilience strategies and, if so, what they are. Also, if we are able to identify resilience strategies can we use this information to enhance the resilience of anaesthetists? Following these interests, this chapter has two main aims: 1) to see if the Resilience Markers Framework can be used to shape interviews to elicit the details of resilience strategies; and 2) to explore resilience strategies in anaesthetists' work.

Background

Safety-I and Safety-II in anaesthesia

Traditional approaches to safety have largely focused on finding the causes of incidents so they can be corrected to reduce the likelihood of recurrence – this has been called Safety-I (Hollnagel, Wears and Braithwaite, 2015). Assumptions that are sometimes associated with these approaches include that the system can be understood, threats can be predicted, and as long as procedures are followed the system should remain within safe boundaries of performance. Here the variability that is introduced by human agents is generally considered a threat to safety, as it brings unpredictability into a stable system.

Safety-II is a different but complementary approach that argues that we should understand how safety and performance is maintained within 'normal' operation (Hollnagel et al., 2015), i.e., not just when things go wrong. Assumptions that can be associated with these approaches include that the system cannot be fully specified and understood, that the system is open to predictable and unpredictable threats, and procedures are often limited in their applicability and detail when faced with real world events. Here the variability that is introduced by human agents is generally considered a positive characteristic, as it provides flexibility in the system to interpret uncertain and ambiguous information and respond to opportunities and threats.

Research into safety in anaesthesia has been predominately carried out by examining the errors made, in keeping with Safety-I. For example, Chopra et al. (1992) and Khan and Hoda (2005) both focus on the direct causes of system failures. Studies of this nature generally seek to understand and reduce the direct failures in the system by either removing a factor that appears to be at fault or adding an additional level of safety. Pronovost, a practising anaesthetist, proposes that there are other factors in addition to understanding the direct failures within the system that are important to safety (Pronovost, Miller and Wachter, 2006). For safety to improve it is important to understand what happens when procedures go correctly too. In anaesthesia there are multiple chances that catastrophe can occur on a daily basis but this is usually prevented. Safety-II proposes to study the successes that prevent failure and maintain performance. Ball and Frerk (2015: 2) recognise the relevance of these approaches to anaesthesia: 'Safety 1 and Safety 2 are not antagonistic, but complementary approaches; Safety 1 investigates the detrimental outliers, while Safety 2 considers the rest, including those who excel.'

The Resilience Markers Framework (RMF)

Hollnagel et al. (2015) recognise 'adjustment' as being a fundamental ability of resilient systems: 'The performance of a system is said to be resilient if it can adjust its functioning prior to, during, or following events (changes, disturbances, and opportunities), and thereby sustain required operations under both expected and unexpected conditions.' The Resilience Markers Framework (RMF) (Furniss et al., 2011b) has been proposed to explore resilience strategies, which account for the ways the system is able to adjust its functioning. These strategies are pertinent to adjustments within normal working and where the system must cope outside its formal design parameters.

The RMF has a three-tier hierarchy at its core: from low-level concrete examples of resilience, to mid-level description of this strategy, to higher-level theory at a markers level. The mid-level descriptive level is the main area of analytic focus, which is divided into four parts: a resilience repertoire, vulnerabilities and opportunities, mode of operation, and enabling conditions. The RMF aims to provide a traceable route of analysis from concrete examples of resilience through to mid-level description, to high-level theory. Furniss et al. (2011b) found that it was common for case studies on resilience to neglect to link empirical observations of resilience to high-level theory. At the same time they found papers on high-level theory that neglected to link their proposals to concrete examples of resilience. It was hoped that this three-tier hierarchy would encourage a better bridge from observations to theory, and back.

The RMF emphasises a focus on behaviours that are 'outside design-basis' (Furniss et al., 2011b). This refers to those behaviours in a system that fall outside of the official operating procedures and what the system is formally designed to do. This includes informal behaviours and workarounds that are intended to avoid unwanted states and to enhance performance. This can be to take advantage of opportunities, to respond to threats, or to compensate for poor behaviour, poor design, poor systems and poor circumstances (Furniss et al., 2011b).

The resilience repertoire represents the library of adjustments a system can operationalise. In some sense the breadth and depth of the repertoire represent the system's ability to respond to different opportunities and threats (including compensation mechanisms). Furniss, Back and Blandford (2010) refer to two mechanisms that can lead to the development of the repertoire: 1) 'Big R' is where a new resilience strategy is created in response to an opportunity or threat; 2) 'little r' is where an existing resilience strategy is adopted and adapted to address a new situation. Whereas Big R can be considered as being more novel and creative, little r could be potentially more powerful between communities and organisations where experts share informal knowledge and resilience strategies to improve performance.

The potential for a system to be able to respond resiliently will also be influenced by the mode of operation and the enabling conditions. The mode of operation provides an opportunity in the framework to capture characteristics of the context that will impact the threat/opportunity and the response, e.g., a first responder's assessment of a mass road traffic accident will impact what resources are enabled to cope with the incident at the roadside and surrounding hospitals. The enabling conditions allow one to capture the specific requirements or precursors that need to be in place to allow a certain response, e.g., a helicopter ambulance can only be sent to a scene if it is available.

Applications of RMF

The RMF was originally developed when trying to identify and describe resilience behaviours exhibited by nuclear power plant control room operator crews, as they tried to negotiate complex scenarios in a simulator (Furniss et al., 2011b). Here we reviewed videos and logs of the crews' performances in the different scenarios, which included expert's commentary on crew performance. To see if the RMF had broader applicability it was tested on observational data gathered from an oncology outpatients unit (Furniss, Back and Blandford, 2011a). Here we noted unwritten and informal practices for maintaining safety and performance. This included behaviour to help nurses monitor their own work, to reduce the likelihood of error (in their own work and potential errors introduced by temporary staff), to reduce complications later in the treatment process and to cope with unexpected scenarios. Work has also started to develop a vocabulary to make resilience strategies more tangible (Furniss et al., 2011) and apply these in health care contexts (Furniss et al., 2014).

Rankin, Lundberg and Woltjer (2011) and Rankin et al. (2014) adapt RMF to incorporate categories from other frameworks, and categories that better reflected their empirical data, which was from a focus group study with people that work in safety-critical domains on the topic 'working near the safety margin'. The resultant framework was the *Strategies Framework*. These more complex, and potentially more comprehensive, developments contrast with Furniss et al. (2011a), who suggested simplifying the framework to just two core categories, i.e., threats/opportunities and strategies. This shows scope to adapt the original RMF depending on the data and the needs of the study.

Method

Rationale for the data collection approach

Our approach to data collection was based on semi-structured interviews, whilst using the categories in RMF to shape those interviews. We decided against observations because although they can be a rich source of information their yields can be low. Furthermore, anaesthetists work in a variety of areas of the hospitals, e.g., when responding to crash calls, which would make observations challenging. Interviews have the potential to engage with a broader range of experiences and data. However, we did not know whether anaesthetists would engage with the topic of these interviews and be able to elicit the sorts of resilience strategies that we were looking for. For example, some of these strategies may be assumed and go almost unnoticed by those who practise them. This made priming and prompting for resilience strategies critical to the success of the data collection.

Data sources for the study

Data was collected at interview sessions with six anaesthetists (one staff grade and five anaesthetic trainees with a minimum of 2 years' practical anaesthetics experience). Recruitment for the study was achieved by one close contact in anaesthesia introducing us to others in their profession. All interviewees were employed in UK hospitals at the time of the study. All had worked in at least four different hospitals in the UK except for one, interviewee 6. Interviewee 6 was a staff-grade doctor who had trained and worked in Ghana prior to working in the UK. Aside from interviewee 6 all interviewees had experience at both smaller district general and major teaching hospitals. They had experience of a range of departments such as obstetrics, paediatrics, and orthopaedic surgery. Ethics approval for this study was granted. The interviewees were not paid for their time.

Data collection

The interviews were held in quiet rooms away from working environments either at participant homes or at the home of the researcher. All sessions took place over a 5-month period as part of the second author's student thesis. The interviews lasted between 30 and 60 minutes. The interviews were held on a one to one basis, only the interviewer and the interviewee were present. The interviews were recorded on a digital recorder and transcribed for analysis. The interviews followed a five-step structure (RMF categories have been highlighted in *italics*):

Step 1 – Introductions and priming for resilience strategies

Before starting the interview the anaesthetists were briefed on the aims of the study. As resilience is not a familiar concept to the interviewees a short video demonstrating and discussing error and resilience was shown.[1] This video allowed us to talk about error and avoiding error in socially acceptable terms, e.g., not forgetting your umbrella before you leave the house, to demonstrate that error need not equate to serious consequences and is part of normal life. This helped build rapport with the interviewee and we explained that the study was essentially interested in how errors are avoided. To elicit meaningful data it was important that the session did not feel like an investigation of the interviewees' previous medical errors.

Step 2 – Contrasting two systems

The initial questions prompt interviewees to reflect on the different demands of their work. By asking them to contrast routine and non-routine aspects of their work we open up the interview to discuss the different tasks, environments and pressures they work within. The aim is to encourage the interviewee to think about situations at work when they are prone to different types of error, so we can discuss strategies to avoid 'regular threats' and 'irregular threats' (Westrum, 2006).

EXAMPLE QUESTIONS

- 1.a. Can you tell me about two different environments, or ways of working in your job?
- 1.b. Can you describe two ways of working, or *modes of operation*, one that is routine and one that you would class as more urgent or unusual?

Step 3 – Probing for differences

This step probes for the features and qualities of the different contexts and ways of working. In emergency situations there will be considerable strains on the system. Time, staff, equipment and resources are likely to push the skills of the anaesthetist to their limit. The aim is to draw out a list of disparities between the two modes that the interviewee thinks may account for potential vulnerabilities.

EXAMPLE QUESTIONS

- 2.a. What are the key differences in the two modes of working?
- 2.b. One way of looking at the differences would be to think of what is available to you in one mode but not in another. Can you tell me about the differences in the *resources and enabling conditions* (i.e., the things you need to enable you to work effectively)?

Step 4 – Identifying vulnerabilities

With the list of differences in the two modes of operation the identification of *vulnerabilities* can start. *Vulnerabilities* may have been mentioned earlier in the interview and if so these can be now discussed in more depth. For example the interviewee may cite that during an emergency call to Accident and Emergency (A&E) Resuscitation (*mode of operation*) there was a lack of time (*enabling conditions and resources*) that led to having to make a quick judgement on a patient without full information (*vulnerability*).

EXAMPLE QUESTION

- 3.a. Can you tell me about the *vulnerabilities* in the different situations?

Step 5 – Resilient repertoires

Following the identification of some of the vulnerabilities the interviewee is asked about the methods used to safeguard against them. A vulnerability may be regular or irregular, and their strategy or method for keeping this in check might be novel or mundane, unique or widespread. It is useful to discuss how the interviewee perceives these strategies and their potential value to the rest of their community.

EXAMPLE QUESTION

* 4.a. Is there anything that you do, you have done or have seen others do that avoids these potential vulnerabilities?

To elicit further resilience strategies that interviewees thought were too minor and insignificant, particularly if they found it hard to recall different strategies, we asked them to recall the video clip at the beginning of the interview and shared the resilience strategies of others', which sparked some inspiration.

Data analysis

The interview transcripts were analysed for the details of explicit examples of resilient behaviour. The interviewees presented two to seven different examples of resilience strategies each. The vulnerabilities the resilience strategies were responding to were analysed using the categories in RMF to add context. It was not mandatory that the vulnerabilities were personally experienced. This study was interested in the way interviewees perceived threats around them, often at many times throughout the day. It was possible that the participant had made a specific error and then adapted their working practice to avoid recurrence. Alternatively, participants may have identified a unique way of protecting themselves from a possible error they have not experienced, or they may have learnt a new 'trick' from a colleague.

Interpretation: from analysis to conclusions

Following data collection the individual resilience strategies were classified in terms of four main themes. These themes represent perceived vulnerabilities and opportunities in anaesthetists' work. Each grouping reflected the activities and goals of the anaesthetists. The groups of resilience strategies were then evaluated in terms of the RMF. We stopped our analysis and interpretation here, as a proof of concept for this exploratory study. However, the resultant strategies could be evaluated in terms of their theoretical contribution (i.e., comparing them with published resilience strategies in the literature) and practical contribution (i.e., what value there would be in sharing these strategies at a team, organisational, or broader community level). For the latter type of contribution, the details of the resilience strategies could be validated and evaluated through member checking, or respondent validation. This is where participants, or a group similar to the participants, review the interpretation of the results to check their accuracy, and as we suggest in this case reflect on their potential value to the community too.

Results

We report seven resilience strategies under four themes: improving the availability of information in patient assessment, preparing for complex procedures, improving access to equipment and drugs in an emergency, and avoiding wrong drug and wrong route.

Improving the availability of information in patient assessment

Resilient strategy 1: Encourage patients to carry prescription and allergy information

Giving an anaesthetic carries risk. Anaesthetists understand that the more information that is known about the patient before the anaesthetic the safer their care will be. In an emergency situation the vulnerabilities can be much higher. Without information about the patient's history the anaesthetist's judgement on drug types or doses becomes very difficult. If the patient is unable to communicate when admitted there is little the doctor can do to gain information quickly. A strategy that Interviewee 5 has implemented on an individual level is to promote drug awareness with incoming patients by telling them about the serious consequences that drug interactions and allergies can have. He advises them that simply carrying prescriptions or information about their allergies in their purses or wallets may help them in the future. For example, he describes the confusion that could be caused by a beta-blocker:

> *R5: If someone comes in collapsed in the middle of the night then we can't phone their GP we don't know what medications they're on. They might have a really slow heartbeat. That might be a problem with their heart or is it because they're on beta-blockers, which will slow down their heart? So you know that if you can increase the amount of information that you have then the decisions that you make can be much better.*

Within the RMF the *vulnerability* is the lack of information that can be conveyed to the anaesthetist at the point of admission. The *mode of operation* might be an emergency where the patient cannot communicate or remember relevant medical information. One *resilience strategy* is to provide an alternative means for clinicians to access the relevant medical information they need. A simple list on a note inside a purse or wallet can be an *enabling condition* for this strategy.

Opportunities to plan ahead

Resilient strategy 2: Preparing work and complex elective surgery

The work of an anaesthetist can involve making many decisions in a short space of time. The types of drugs, the dosage and the type of anaesthetic are all critical decisions of this nature. Two of the interviewees explained that being provided with details of their rota and scheduled elective surgeries gave them the opportunity to plan ahead. This included mentally preparing and rehearsing potential pitfalls for different contexts, e.g., whether they are doing emergencies or obstetrics. It also included more detailed planning of complex surgery:

> R6: *Sometimes with an elective case you know what you are doing tomorrow. And you have the whole night to go and read about it and prepare it and sort of strategize.*

Both interviewees 1 and 6 were aware that making a number of different decisions in a short space of time makes them *vulnerable* to error. Information on rotas and elective surgery prior to the working shift can be considered an *enabling condition*. The *strategy* can be described as 'taking the time for mental preparation'. The behaviour presented can be defined at the broader *marker* level as 'preparation'.

Improving access to equipment and drugs in an emergency

Resilient strategy 3: Take what you need with you

When responding to an emergency call the anaesthetist leaves the familiar environment of the anaesthetic department and is required to work in a less familiar workspace. An emergency call could direct the anaesthetists to the A&E Resuscitation area, a labour ward or other parts of the hospital. This presents a new set of vulnerabilities. An anaesthetist will often work in these areas without the dedicated assistants (Operating Department Practitioners (ODPs)) that they are used to working with. Instead they will work with nurses and other staff that may not understand the requirements of their job. Interviewee 3 gave this example from working in an A&E Resuscitation room:

R3: I had an example the other day where I went down and basically asked for a face mask for a patient to give bag-masked ventilation and they only had one size of face mask. I said I need a smaller one and they all said we've only got one size and they thought that I was crazy for suggesting that we had more than one size of face mask. And I said 'Well you know this is ridiculous. If we had a small child or an adult we are going to need a different face mask.' Everyone was adamant that they didn't have different sized face masks and it took me about 10 minutes to get a smaller face mask.

Delays in dealing with a patient were cited by all of the interviewees as being a concern. Aside from unsympathetic assisting staff there are issues around the availabilities of drugs and equipment. Anaesthetists use certain types of drugs in an emergency situation: vasopressors, muscle relaxants as well as anaesthetic drugs. These drugs, in theory, are present throughout the hospital. In practice they are often difficult to locate or are locked away without a key holder present. A strategy that the some of the interviewees have used is to carry emergency drugs with them. Interviewee 2 describes her strategy:

R2: What has happened, what I have done in the past, is if I know I'm going to be called down to A&E Resuscitation, that I take my drugs with me from theatres, that's one thing I've done in the past. Drugs that I need, to use, to put people to sleep very quickly. So those drugs are Thiopentone, Suxamethonium and Propofol. Those would be the main drugs I would take. I take them with me because then I have immediate access to them. I'm sure they would be somewhere there in A&E but again, as I said before, when you need these drugs you need them quickly. And you can't wait five to ten minutes for someone to go and get them.

Interviewees 1, 2 and 5 were aware of the *vulnerability* that the departments they were being called to had 'potentially limited supplies of drugs and equipment' that they needed to carry out their work. The *mode of action* can be described as 'an emergency', which needed urgent action. The *strategy* employed can be described as 'taking drugs and equipment to emergency call' to avoid delays. In some hospitals respondents referred to this strategy being more formalised as anaesthetists had an 'emergency drug pack' that they took from the anaesthetic department, which has all the mediations they need for an emergency.

Resilient strategy 4: Differentiating keys and cupboards using stickers

Interviewee 3 implemented a different strategy that aimed to reduce the time taken to acquire drugs and equipment during an emergency. He found that when he and his colleagues were given the keys for the emergency drug cupboards in the A&E resuscitation room finding the right key for the right door took too long:

> R3: You've got four keys. You don't know which one it is, they all look the same for these cupboards. [pause] It does happen when you're given the keys for A&E or something like that. They give you the keys, you try and get the drugs out and you don't know which key it is. It just wastes loads of time.

Interviewee 3 had coded the keys and cupboards so that all anaesthetists in the department were able to locate the drugs quicker than before. The *vulnerability* presented was 'difficulty in locating supplies of drugs and equipment'. The *mode of action* can be described as 'an emergency'. The *strategy* employed can be described as a differentiation strategy: 'creating a colour reference system' to match keys with doors. At the broader marker level this strategy has some relation to 'strategies that maximise information extraction', which was featured in Furniss et al. (2011b).

Avoiding wrong drug and wrong route

Resilient strategy 5: Differentiating drugs using coloured stickers

The Royal College of Anaesthetists' guidelines recommend using International Colour Consortium Specification (ICCS) colour-coded stickers to tell drug types apart, e.g., vasopressors have purple labels and narcotics have blue labels. As all interviewees adhered to these guidelines the 'strategy' was of reduced interest to this research. However, how the interviewees combined the ICCS colour-coding with additional methods or used the colour-coding under different conditions presented more interesting data. For example, Interviewee 3 used the ICCS to colour-code the keys and cupboard doors so the right cupboard could be located for certain drugs, and the right key could be easily located for that cupboard.

Interviewee 6 used stickers as an additional marker on infusion bags he was using. Existing policy required infusion bags to be marked with written descriptions of what they contain. However, Interviewee 6 added ICCS colour-coding stickers so that the drug types the infusion bag contained was more salient from different angles and distances.

Interviewees understood that infusion bags look similar and connecting the wrong one is a *vulnerability* in the system. This *strategy* can be described as 'creating a secondary indicator'. At a broader *marker* level this strategy can be considered within 'strategies that maximise information extraction'. Fellow anaesthetists, ODPs and nurses were all able to benefit from this strategy.

Resilient strategy 6: Differentiating drugs using different sized syringes

There are different types of vasopressor drugs, which have different effects, so they are not interchangeable. Interviewee 3 discussed how he uses a different sized syringe for the different vasopressor drugs. This reduces the likelihood of picking up the wrong syringe when they look similar:

> *R3: The ones that we commonly use are Ephedrine, Phenylephrine and Metaraminol. Phenylephrine and Ephedrine are quite similar names. They do have different actions. And potentially picking up, if someone had a slow heart rate the Phenylephrine and Metaraminol can cause that to go a bit slower while bringing the blood pressure up (it) can cause problems so that giving that in that situation and basically you can see if they are all in the same sized syringe. I used to put Metaraminol in a 20 ml syringe and Phenylephrine and Ephedrine in a 10 ml syringe which made the numbers quite nice, and basically there is a situation where you could pick up a purple 10 ml syringe and it could have been the wrong one quite easily. So I've started always putting my Ephedrine in a 5 ml syringe. So it's double the dose, you know double the concentration that you use, to put Phenyleprine in a 10 ml syringe and Metaraminol in a 20. So whenever I draw it up I do that and I know that with the size of the syringe and the sticker, I'll always be getting the right one.*

Using a combination of syringe size and colour strengthens the strategy. Interviewee 1 also referred to this strategy:

> *R1: And the other thing that I kind of started to say was that certain drugs go in certain syringes so that in the heat of the moment I don't grab some thing and give it and don't know what it is.*

Both interviewees were concerned with identical syringes containing different drugs within the same class. The drugs look similar but have different effects on the patients. This creates a *vulnerability* in the system. The *strategy* is a differentiation strategy. Having different size syringes available for use like this is an *enabling condition*.

Resilient strategy 7: Reducing wrong route connections

Interviewee 5 gave an example of a strategy he used to reduce the likelihood of giving a drug prepared for intravenous route (into a vein) through an epidural (into the spine). The consequences of this are grave and can cause significant damage or death to a patient. Interviewee 5 told how he left the syringe attached while not in use so that delivering the wrong drug would be impossible without removing the correct drug:

> *R5: With the epidural, the drugs that you draw up they are in the same syringes as you'd use for intravenous. And with the epidural, when you're sort of topping someone up for a caesarean section you don't give the whole lot in one go. Because that would be dangerous. You give a little bit, wait, you give another little bit, you wait, you give another little bit, wait. Now if you physically detach the syringe each time that you give a little bit then you could attach, you could grab hold of the wrong syringe and give something and potentially accidentally pick up that syringe and give it IV. To avoid that I only give say 5 ml at a time but I leave the one syringe attached. All the time. So I can't accidentally give that syringe anywhere else. And I can't actually give any other syringe to there.*

This respondent recognised the *vulnerability* of connecting an intravenous drug to the epidural route. The *strategy* can be defined as 'reducing complexity and decision making' as different syringes were not continuously detached and reattached, which could increase the likelihood of attaching to the wrong route. Of course, having a syringe constantly attached to a line, with more drug than is necessary, could have its own risks and vulnerabilities.

Discussion

This chapter had two main aims: 1) to see if the Resilience Markers Framework can be used to shape interviews to elicit the details of resilience strategies; and 2) to explore resilience strategies in anaesthetists' work. We address each in turn.

The interviews were fairly straightforward to develop using the RMF, which suggests it's a strong framework for this purpose. Others could use and adapt our five-stage interview script to investigate resilience strategies of professionals in other contexts. Stage 1 used a video to help introduce resilience strategies and show that resilience strategies are part of the everyday battle against error and for the improvement of performance. This proved to be an effective priming tool. The subsequent four stages of the interview covered contrasting two systems, probing for differences, identifying vulnerabilities and the resilience repertoire. All four stages allowed the interviewees to reflect on their own working practices, the different environments and conditions, the vulnerabilities that are inherent in the work and the opportunities they have to protect the patients and themselves from error. Prompting for resilience strategies included reminding them of the video that seemed to help them recall those minor, perhaps seemingly insignificant, actions that provide resilience. Strategies elicited from previous interviews were also discussed to get their opinion on them and to encourage further ideas. Due to the positive nature of the discussion, this framework could allow for more discourse around error, avoiding error and improving performance in a non-threatening way.

Resilience strategies are relevant for anaesthetists' work. We have identified seven strategies that include improving patient assessments in the longer term, improving the availability of drugs and equipment in emergency situations, and planning ahead. This was a small exploratory study, with six anaesthetists, and all were able to report some resilience strategies. Those that had worked in more than one hospital referred to different practices, e.g., it was reported that only some hospitals had an 'emergency drug pack' that the anaesthetist could take with them from the anaesthetic department to emergencies in other areas of the hospital. This shows that even more formal resilience strategies are not disseminated widely. We see recognising these strategies as the first step before further work that should ask the community to evaluate them and disseminate them more widely.

Acknowledgements

We would like to thank the respondents for the time they gave to this study. This work was supported by the UK Engineering and Physical Sciences Research Council [EP/G059063/1].

Note

1 www.youtube.com/watch?v=Hm7k0TRaPHI

Chapter 8

Promoting resilience in the maternity services

Cathrine Heggelund and Siri Wiig

Introduction

Health care services today are characterised by the increasing use of sophisticated equipment, new technology and demands for efficient treatment and better care for a steadily rising number of patients. Longer life expectancy is leading to more complex clinical challenges and future health care services will have to respond to patients' greater demands and expectations, but with fewer available health care professionals. At the same time, health care services need to be performed safely and in a cost-effective way (Hollnagel, Braithwaite and Wears, 2013a). If health care services are to be sustainable and safe, they must be able to adapt to environmental changes and quickly identify solutions to cope with unexpected events – even with limited staff and resources (European Commission, 2014).

Over the last few years, western countries have established a stronger organisation for quality and safety in health care through reforms, more intensive research, and improved quality and safety standards for primary and specialised health care services (Hjort, 2007; Aase, 2010; Jha et al., 2010). The increased focus is apparent in a series of political and academic national and international guidelines (Ministry of Healthcare Services, 2010–2011; WHO, 2004). The EU has established its own patient safety recommendations and the World Health Organization established a separate patient safety programme in 2004 (WHO, 2004).

Patient safety is gaining increasing attention in Norway and is visible in numerous strategic documents, including the National Strategy for Quality Improvement in the Health and Social Services (Directorate of Health, 2005), the National Health Plan (Ministry of Healthcare Services, 2010–2011; 2014–2015), and in annual reports on quality and safety to the parliament in health services (Ministry of Healthcare Services, 2012–2013; 2014–2015). In addition, patient safety has been codified in several laws and regulations in the health care system.

However, despite the enormous efforts and actions taken to reduce adverse events and improve patient safety, challenges remain in relation to patient safety and the number of adverse events is still too high (Braithwaite, Wears and Hollnagel, 2015; Vincent and Amalberti, 2016; Pronovost et al., 2015). The patient safety research so far has had a reactive focus (Laugaland, Aase and Waring, 2014). Several studies have focused on adverse events and why these occur. This research has generated knowledge about influencing factors that contribute to such events. In contrast, there has been little research on factors and mechanisms of importance for enabling organisations to anticipate risk, respond to variations in performance, monitor organisational processes and outcomes and learn from both positive and negative risk events. In other words, there is need for more knowledge about the mechanisms that create resilience in health care systems (Braithwaite et al., 2015; Hollnagel 2013b).

Aim and research question

Resilience research has centred on resilience theory: how resilience performance occurs. There is now a need for attention to the implementation and evaluation of resilience in practice (Righi, Saurin and Wachs, 2015). This study explores the mechanisms shaping resilience in maternity services in two Norwegian hospitals. By applying the four cornerstones of resilience, we will identify what mechanisms maternity wards use for anticipation, monitoring, response and learning in their daily work and discuss how others can adopt these mechanisms to promote resilience. We then compare the two maternity wards to explore patterns of importance across different contexts. The following research question guided our study: What mechanisms do maternity wards use for anticipation, monitoring, response and learning in their daily work?

Method

Rationale and context

This study is part of the EU FP7 QUASER project (Quality and Safety in European Union Hospitals). It is based on the data set obtained through the Norwegian arm of the QUASER study according to an internationally accepted study protocol of a case study of two Norwegian hospitals (Robert et al., 2011).

The design is a comparative multi-level case study (macro, meso, micro level) of an urban university hospital and a rural hospital (Robert et al., 2011). A case study examines the empirical material of which phenomena cannot be detached from the context and is suitable when 'how' or 'why' questions are posed and when dealing with a temporary set of events (Yin, 2009). A case study can therefore be an empirical study that explores a phenomenon within a particular context and period. One cannot study the phenomenon in isolation from context or time (Yin, 2009).

The Norwegian QUASER team selected two hospitals based on common criteria across all five participating countries (Burnett et al., 2013). The criteria were based on five national quality indicators and the participating hospitals in the QUASER project were ranked on the basis of these. One of the Norwegian hospitals was among the top performing hospitals in the country; the other was in a developmental phase (Wiig et al., 2013). Table 8.1 presents key information about the two hospitals.

Hospital 1 had one of the smallest women's health clinics in Norway. The hospital had a maternity ward with expertise and equipment to receive mothers with moderate risk issues, according to the selection criteria described in the 'Quality Requirements for Maternity Care' (Directorate of Health, 2010). In this maternity ward, there were approximately 1,000 births a year.

In 2010, the maternity and the gynaecology departments in hospital 1 were merged. The gynaecology department originally had 19 beds. Based on diagnoses, three of these beds were transferred to the maternity ward, and the remaining beds were transferred to other departments. Therefore, the maternity ward at hospital 1 was a combined maternity/post-natal/gynaecology ward with 19 beds and associated outpatient clinics. The ward consisted of four delivery rooms and three beds were reserved for gynaecology patients. The department staff consisted of gynaecology nurses, midwives and nurses under the supervision of the head midwife.

Hospital 2 was a large university and teaching hospital, which means that its main responsibilities, in addition to patient care, were research, education of health professionals, and training of patients and relatives. Hospital 2 was characterised by a large medical research environment.

The maternity section at hospital 2 had an observation ward, a combined maternity/post-natal ward (midwife-led ward), a maternity ward with responsibility for births requiring follow-up, monitoring and treatment, and two post-natal wards. The maternity section had its own operating room and eight delivery rooms. The staff consisted of doctors, midwives and pediatric nurses.

Table 8.1 Hospital information

	Hospital 1	Hospital 2
Localisation	Rural town in Norway	Large town in Norway
Number of staff	About 2,200 employees	About 11,500 employees
Range, quality indicators	Top layer	Development phase
Maternity department, number of births a year	About 1,000	About 5,000

Data sources

This study is based on data from the QUASER study in Norway. The data sources consist of qualitative interviews, focus group interviews, field notes from observations (meso and micro level) and analysis of national documents (macro level).

The total data set from the Norwegian QUASER study comprised 99 interviews, field notes from 45 hours of observation, and focus group interviews collected in 2011–2012. At the micro level, interviews were conducted in the maternity ward at each hospital and in one cancer ward at hospital 1 (Wiig et al., 2013). In this chapter, we apply data from the two maternity wards, one at each hospital.

All interviews at the micro level of maternity wards were included in the study. This covers micro level staff from new graduates to seasoned midwives, nurses and doctors. To supplement the micro level data, a selection of interviews from senior managers at meso level was included. Criteria for selection of meso level interviews was that the informant had to be affiliated with the maternity ward and/or have a position in patient safety. In addition, six field notes from shadowing of health care personnel were included in our data material. Shadowing can describe organisational factors and details that are difficult to discover in the interviews. Table 8.2 summarises the interviews included in this study.

Data collection

The data set applied in this chapter is based on qualitative interviews (38), observation and shadowing of staff (35 hours) and document analysis. Each interview lasted approximately 1.5 hours. The QUASER study obtained ethical approval from the Norwegian Social Science Data Services (NSD), (NSD reference: 26636) (Wiig et al., 2013; Robert et al., 2011).

The informants were recruited by a site manager in each hospital and the snowball method was used when needed. All informants received information about the purpose of the study and signed a written informed consent.

The interviews and observations were conducted according to an agreed-upon interview and observation guide in the consortium. The interview guides used at the micro level focused on the role, responsibilities, teamwork, culture, management, quality improvement, IT systems and how the participants experience work at their unit. The observation guide followed similar topics (Wiig et al., 2013; Robert et al., 2011). All interviews were conducted in case hospitals by the Norwegian QUASER team. The interviews were tape recorded and later transcribed. All transcribed material has been kept confidential. All informants were assigned individual codes to ensure anonymity and data was stored in locked cabinets or on PCs in secured files against unauthorised access.

Table 8.2 Overview of data sources

	Hospital 1	Hospital 2
Micro level Maternity ward	Constituted head of department Department nurse, gynaecology polyclinic 2 pediatric nurses Professional development midwife Gynaecology nurse 3 midwives Head midwifeSenior consultant Focus group interview: (Gynaecology nurse, midwife and pediatric nurse)	8 midwives Pediatric nurse Senior consultant Doctor in specialisation 2 department midwives 2 assistant department midwives Professional development midwife Focus group interview: (4 midwives, assistant resident, 2 senior consultants)
Meso level	Clinic director, Surgery Constituted head of Women's clinic Director of development Medical director Advisor	Clinic director Medical director Head of secretariat for patient safety Head of department for data management
Shadowing of staff	3 (Midwife, 'duty officer nurse', gynaecology nurse)	3 (Doctor, doctor in specialisation, coordinating midwife)
Total	17 interviews + 3 shadowing	21 interviews + 3 shadowing

The QUASER team completed two rounds of interviews, approximately at 6-month intervals, during which some of the informants were re-interviewed (Roberts et al., 2011). In this way, misunderstandings could be resolved and missing data from the interview in round 1 could be collected. In addition, the second round of data collection served as a validation process and member check.

Data analysis

This study applies theory-driven content analysis. This analytical approach means that the text was sorted according to predetermined categories that originate in a particular theory (Malterud, 2011). The four cornerstones of resilience – anticipation, monitoring, learning and response – are applied as the theoretical framework for this analysis.

As the first step, two separate within-case analyses were performed, one from each hospital. The data was analysed according to the four cornerstones. In order to analyse patterns across the hospitals, the results from the two maternity wards were later put into an across-case analysis. The aim of the across-case comparison was to identify similarities and differences in mechanisms and factors that contribute to resilience in maternity wards in hospitals that deliver maternity services in different contextual settings.

The in-depth content analysis followed Malterud's (2011) five-step method for text condensation. The following five steps summarise the process of the content analysis:

1   To form the overall impression of the material.
2   Identify meaning units in the text.
3   Densification of the text to a brief summary.
4   Coding processes.
5   Result description.

The first step consisted of becoming familiar with the material to form an overall impression. At this level, Malterud (2011) points out that the pre-understanding and the theoretical framework should be temporarily set aside. This way, we could consider the material in the most transparent and independent way.

All interviews were thoroughly read and it was then, on the basis of material relevance, that we made a choice of micro-systems (to not include the cancer care micro system) as well as interviews from meso level that would be used to form the basis for this study. The material was revised again and notes of relevant statements in the text were simulta-neously taken. These themes were reduced later in the analytical process when several overlapping themes emerged. This happened after discus-sions between the researchers and several readings of the data material. Malterud (2011) highlights the importance of discussing the material with other researchers because discussions can reveal details and nuances. It is an important step in validating the results.

As this study contains a theory-driven analysis according to resilience in health care, the four cornerstones of resilience formed the basis for the formulation of the categories. Then we broadened our search for materials that contained elements of these four topics. Many themes and categories emerged along the way and hence a need arose for further concretisation. Tables systematised the data. Table 8.3 shows an example of the analytical process starting with quote, meaning unit, coding and categorisation, according to Malterud (2011).

All categories from the interviews were collected and grouped for content analysis. Both micro systems were analysed separately. In retro-spect, a cross-case analysis of factors that promoted anticipation, mon-itoring, learning and response was carried out. The contextual aspects of the micro systems were included in the interpretation of the results, such as the size of the maternity wards and the organisation. Categories and key factors that emerged in each hospital were incorporated into a new table, which illustrates both similarities and differences, in addition to the robustness of factors that emerged at each of maternity wards (Yin, 2009).

Table 8.3 Example of analytical process

Quote	Meaning unit	Code	Category
'But we always have two midwives when a woman starts to push and the head appears [. . .]. It has to do with rupture prevention, then you have someone to assist. There is a routine that says there should be two midwives during delivery to prevent rupture.'	Routine that there should be two midwives during delivery to prevent rupture.	Prevent	Anticipate
'There's always someone you work with that you establish a special/good tone with . . . some is a little bit more 'on' – and will do too much, while others have a more relaxed attitude – it varies. It is all about personality and professional approach.'	Some people work better together than others. Personality and professional approach.	Cooperation	Respond

To understand macro level expectations for a resilient maternity care, we analysed two strategic documents that are of particular importance for safe patient care in the Norwegian maternity context. White Paper 12 (Ministry of Healthcare Services, 2008–2009), 'A joyous occasion', describes what a comprehensive, coordinated and safe maternity care should include and how it should be achieved. Responsibilities and cooperation between the primary and acute care services are listed.

The White paper proposes several areas for improvement and reiterates that the responsibility for birth and maternity care rests with the regional health authorities. The government proposes that regional health authorities and local health trusts must prepare a multi-annual, integrated and locally adapted plan for maternity care in the region. The aim is to ensure sound capacity, resources and expertise and to provide decentralised, diversified and predictable maternity services.

'"Safe maternity care". Quality requirements for maternity units' (Directorate of Health, 2010) describe the legislation and quality requirements for maternity services. It gives the requirements for the organisation of maternity units, staffing and expertise, quality, and improvement, as well as guidelines for a risk-based selection of mothers. The system should have procedures that contribute to good communication, cooperation, and allocation of responsibility (Directorate of Health, 2010).

These macro level strategic documents offer important guidelines for the organising of maternity services and provide guidance for what is required from the institutions involved in this study. These requirements pertain to organisation, competence, training, instruction and staffing. They form part of the contextual conditions of the maternity wards in our study.

Interpretation

In this section, we will present our interpretation of the findings from both case hospitals according to the four cornerstones of resilience: anticipation, monitoring, learning and response.

Anticipation

Despite major differences in the ways of organising, our interpretation of the results showed many common factors of significance for resilience in maternity wards. Table 8.4 summarises the main mechanisms used to anticipate potential risks.

Both maternity wards used written guidelines to strengthen the ability to anticipate and respond. For the same reasons, medical devices were frequently controlled according to a checklist.

The quality handbooks contained updated procedures and guidelines applicable to the wards. All informants were familiar with the ward's quality handbooks and these were available to all staff, on the hospitals' intranet.

> The electronic quality handbook is at any time up on one of computers on duty room- methods are there [. . .]. So with one keystroke you get up the procedure. (Assistant department midwife, hospital 2)

Table 8.4 Anticipation mechanisms

Mechanisms used to anticipate	Hospital 1	Hospital 2
Procedures	Quality handbook. High degree of use.	Quality handbook. High degree of use.
Organising	Duty-officer – having oversight.	Coordination midwife – having oversight. Coordination centre – providing resources and competence.
Forums to discuss safety at department level.	Not specified – occurs in staff meetings.	Not specified – occurs in staff meetings.
Medical equipment/ Preparedness	Fixed control routines.	Fixed control routines.
Team composition	Carefully planned based on competence.	Carefully planned based on competence.

Both maternity wards had control routines and checklists for the medical equipment. The managers were responsible for ensuring that all employees were familiar with and complied with the procedures relating to these control routines.

The wards were organised respectively with a 'duty officer' and a coordinating midwife. Both held a clinical leadership role and were a key factor in predicting birth progress, and contributed to enhanced flexibility and redundancy on the shifts. Team composition was carefully planned. The managers and the coordination midwives composed teams based on competence, experience, educational needs and allocated responsibility in order to cultivate each team's ability to anticipate. They also used routines, simulation and controls of medical equipment to prepare the emergency room.

The coordinating midwife and 'duty officer' were always experienced midwifes who led the ward on a shift. Their main tasks covered delegating tasks, calling for extra resources, making telephone calls and being available for other midwives for counselling and support. A main characteristic of the role was having the big picture of the ward's activities, births, personnel resources and familiarity with emergency procedures and calling for help. The interpretation of the results showed that the coordinating midwife and 'duty officer' held an important position among the midwives:

> And the fact that there are always six midwives, meaning in my work, six midwives, where one of them is the main supervisor who has an overview. It is one of the veterans [...] I think that's a very good system. Then there is always someone who has an overview and knows where the doctors are located and can manage or help when we need medical attention [...]. Yes, a person who assists in the department. (Midwife, hospital 1)

Hospital 2 was a university hospital with a large women's clinic. It was organised differently from the maternity services at hospital 1. Hospital 2 had established a coordination centre consisting of a pool of employees with no departmental affiliation, but they were allocated according to a resource needs principle. The coordination centre improved the ability to allocate resources according to a risk-based approach.

Neither of the maternity wards held regular meetings with patient safety with a specific topic, but it was occasionally brought up and discussed in staff meetings.

Monitoring

The hospitals demonstrated several common factors used to monitor safety and performance. In Table 8.5 we list the monitoring tools.

Table 8.5 Monitoring tools

Tools used to monitor	Hospital 1	Hospital 2
Quality indicators	High focus	High focus
Screening	Yes	Yes
Using medical equipment	CTG	CTG+STAN

Both hospitals focused on local and national quality indicators and used screening according to national guidelines. Proactive and reactive indicators were important for the ability to anticipate and respond.

The hospitals used the indicators as a learning tool and as a basis for evaluation of their practice. In addition, the indicators served as a basis for comparison, both internally and against other hospitals in their monitoring of everyday risks. The maternity ward at hospital 1 used a separate computer system, as well as national quality indicators, to monitor several parameters, such as the number of forceps or vacuum deliveries, use of epidurals, the number of elective and emergency caesarean section, rates of incontinence and bleeding, and rupture rates. These statistics were available for all the employees. The staff could monitor the statistics in the hospital's monitoring programmes, where they could find both the hospitals' internally developed indicators in addition to the national quality indicators. This was a well-known and frequently used tool:

> Number of ruptures that harm women during births, from day to day, the value of it is that all employees can access the statistics and press 'my department' and then they see. Then, when we have focused on ruptures, for example, we've reduced number of ruptures .... Except that, we have done nothing specific [...], but we've talked about it. So, we kind of had academic discussions. Because it comes up, and that is important. They all monitor from week to week. Yes, and so far this year we have only had 1% and 1.5%, and it's awesome, and it creates, in a sense, an enthusiasm and drive to be good at it. (Constituted Head of Department, hospital 1)

In both hospitals, quality and safety were a line management responsibility, but hospital 2 established a patient safety support unit to contribute to the implementation of tools to improve patient safety. The patient safety unit supported hospital managers, and drew attention to risky aspects of treatment, care or system aspects that could harm the patient. The unit's work was partly based on reviewing adverse events and near misses. Results from this monitoring were fed back to the managers and fostered their anticipation of new trends.

As a university hospital, hospital 2 possessed more specialised expertise and therefore used ST analysis (STAN) as a monitoring tool, as well as cardiotocography (CTG), which was also applied in hospital 1. CTG and STAN are frequently tools used to monitor high-risk pregnancies where women had risk factors such as diabetes or epilepsy. Following the foetal heartbeat could provide important information about the foetus's condition and its ability to survive birth. To anticipate possible risk situations linked to birth, the maternity wards used CTG-surveillance of foetuses. Procedures for when CTG were recommended were described in the quality handbook. In hospital 1, CTG monitoring pictures were only available at delivery rooms. By moving the CTG monitoring from the duty room to delivery room, the midwives at hospital 1 had to call a doctor into the delivery room for advice.

At hospital 2, the monitoring was available in the delivery rooms, the duty room and from the doctors' offices. Field notes from the duty room showed that this room served as a meeting point for all professions. The staff frequently discussed patient-related challenges.

Both hospitals used screening according to national guidelines.

Learning

The two hospitals showed both differences and similarities in their ways of learning, as illustrated in Table 8.6.

Both maternity wards used practical exercises and simulation as an important mechanism for learning and to improve the ability to respond; hospital 2 to a greater extent than hospital 1.

At hospital 2, they focused on specific topics and techniques and their professional development midwife was highlighted as the driving force for updating and improving the routines and practical training when there was free time.

> When we have some spare time, our professional development midwife is clever and says that 'now we go and practise stuck shoulders' or 'now we go through this and that'. (Midwife, hospital 2)

Table 8.6 Learning mechanisms

Mechanisms of relevance for learning	Hospital 1	Hospital 2
Practice/simulation	Moderate focus	High focus
Training (new employees)	Good	Mediocre
Openness	High degree	Varying degree
Culture of blame	Low degree	Some degree
Quality improvement projects	Yes	Yes
Job rotation	No, only one department	To some degree

The management encouraged practical sessions based on previous incidents where they experienced improvement needs in relation to teamwork and expertise. The results showed that the informants called for more multidisciplinary training with the aim of enhancing the quality of collaboration and communication. The maternity ward therefore scheduled two days a year where doctors and midwives participated in interdisciplinary seminars.

The maternity ward at hospital 1 also organised interdisciplinary seminars. They held regular courses in emergency situations such as seizures, bleeding, acutely ill newborns and breech births. These courses were held three times a year and the managers' goal was for all employees to attend these courses at least once a year. The interpretation of the results showed that these courses were important training of skills and communication.

Both hospitals used quality improvement projects as a learning tool, and to improve services. Job-rotation between wards in the maternity section was a mechanism applied by hospital 2 to increase competence and flexibility. Job-rotation was not applicable for hospital 1 since the maternity ward consisted of only one department.

The maternity ward at hospital 1 showed greater attention to educational activities and a higher degree of openness characterising the organisation than hospital 2 did. Hospital 1 provided training for temporary staff, and the staff expressed a welcoming attitude towards new employees and temporary workers and highlighted the importance of learning from one another. Hospital 1 used an online educational programme that made several training programmes available for its staff.

The staff at hospital 2 expressed a variety of opinions of the education and introduction programme at the maternity ward. Several spoke of the lack of organised training and complained that they were given too much responsibility during their first shift. The support from the staff varied. The management admitted that the introduction programme was somewhat deficient:

> We've trained all the time, but we have not set aside time specifically for training, so one often starts his first duty as part of the basic staffing. So it's certainly something we think is inadequate; there's no time for training the new employees. (Department midwife, hospital 2)

The interpretation of the results from the maternity ward at Hospital 1 showed an open climate where it was easy to discuss both satisfactory and unsatisfactory performance. In contrast, there was greater variation in informants' responses around the transparency and correspondingly more blaming of individuals at hospital 2. At both hospitals, adverse events were sometimes addressed at staff meetings and used as a learning source. Hospital 1 showed greater willingness to report incidents than hospital 2. Despite attempts to create a regular routine for the management of adverse events, there was no permanent structure in hospital 2. Adverse events were discussed to varying degrees, and some informants stated that reprimands were issued.

Interpretation of the results from hospital 1 showed a systematic approach to the management of adverse events and near-misses. A culture of openness characterised hospital 1 and the openness constituted an important mechanism to promote learning and increased ability to anticipate and respond to similar events in the future.

> *We gather when we're done ... as soon as possible so we go through, midwife and nurse and gynaecologist and talk through. What do we think happened? It is very useful [...]. Did we overlook something? We're going through the registration process if we have it, for example, and look. Yes CTG. What time did we begin to discover that perhaps this child was stressed? Or, haven't we seen anything of it? (Midwife, hospital 1)*

Responding

Results from this study showed significant differences related to the factors and mechanisms for response. This is summarised in Table 8.7.

To enhance flexibility and ensure available resources the coordinating centre at hospital 2 was an important mechanism. The coordination centre consisted of several midwives and nurses prepared to start the day at one ward but could be reallocated to another if capacity or expertise needs indicated that. The coordination centre provided an adaptive and risk-based allocation of resources and competence to solve acute situations safely. This feature increased the flexibility of the maternity services.

Table 8.7 Response mechanisms

Mechanisms of importance for ability to respond	Hospital 1	Hospital 2
Available resources/flexibility	Joint (maternity, post-natal, gyn)	Coordination centre
Collegial support	High degree	Varying
Professional support	High degree	High degree
Quality of collaboration between professional groups	Good	Varying
Organisational learning	High degree both informal and informal	Mediocre, more was desirable
Experience of IT	Unsatisfactory	Unsatisfactory

It must feel safe to be in such a type of emergency department. It must feel safe and not irresponsible. Of course, there are times where it is too busy, but then we have a coordination midwife that we contact if necessary. She can provide additional help. (Midwife, hospital 2)

Since hospital 1 was a small hospital where maternity and gynaecology services were in the same ward, joint personnel resources were used to ensure flexibility.

The two hospitals differed in cohesion, support and cooperation. Hospital 1 demonstrated a high degree of professional and collegial support, and good collaboration climate across professions. All informants at hospital 1 described the cooperation across professions as good and built on mutual respect. The midwives and physicians worked closely together and supported each other during the day. In addition, the informants pointed out the advantages of being a small unit. The organisational size contributed to close cooperation, in-depth knowledge about colleagues and their competence and resulted in a collective sense of responsibility.

Hospital 2 showed a range of collegial support and quality in the interprofessional collaboration. Several informants observed that the amount of support and cooperation depended on the individuals:

You know that someone is more helpful than others, so you quickly learn who will hang up the phone and who will answer. It is unfortunately a fact [...]. So, people must be able to tie their shoelaces themselves before coming here. (Doctor in specialisation at hospital 2)

Overall, the midwives at hospital 2 reported good within-professional relationships, but results showed challenges related to collegial support across professions.

Maternity services at both hospitals experienced IT systems as unsatisfactory, mainly because both physicians and midwives documented their notes in different programmes. The programmes were not connected. In addition, midwives had to use several other programmes to document the condition of mother and baby.

Discussion

In this study, we have explored maternity services in two hospitals with great differences in size and organisation. However, the maternity services shared several resilience mechanisms. By identifying and analysing the similarities and differences, we gained insight into resilience factors used in everyday work. These factors illustrate the content of the four cornerstones and increased knowledge about factors that promote anticipation, monitoring, learning and response in the maternity services.

In order to implement resilience in practice, we argue that organisations need to develop capacity to diagnose themselves and understand factors that contributes to both success and failure. Based on our results, flexible organising is a key resilience mechanism in order to provide a risk-based maternity service in which the most competent and experienced personnel are assigned to high-risk patients. Cultural factors are also of great importance to the promotion of resilience. Practitioners who promote resilience in their contextual settings should focus attention on building a culture of openness, support, communication, cohesion and trust.

Work plans at both hospitals were prepared with two aspects in mind – to create learning situations for new employees and to provide safe services. The services solved this by mixing experienced and inexperienced personnel in a deliberate work plan process. Several studies have shown that knowledge and experience improve the ability to predict adverse events (Cuvelier and Falzon, 2008; Vincent, 2010). This was an important mechanism to maintain the ability to anticipate and respond. The lesson learned is to have a long-term perspective on work and shift plans, where the fundamental idea is promoting learning and improvement among staff, not just to get through the shift. We also believe that having a buffer of staff familiar with the services is essential to promote a resilient performance in times of, for example, sick leave.

The 'duty officer' and the coordinating midwife were instrumental in delivering safe maternity services. Having an experienced person with the overview helped to create capacity for other midwives and that person was a liaison between groups. In addition, midwives rotating in this position were frequently used as a 'second opinion' in the delivery rooms. This type of organisational redundancy depends on both the availability of personnel and an open and supportive culture, and thereby helps the micro system to respond. The implications for others trying to foster resilience could be to give priority to allocate resources to a function with the main task of securing oversight and status on key tasks involved in service provision.

Both maternity wards relied on quality handbooks and procedures. Previous research on resilience in health care emphasises that standardisation alone will not be a solution as system is too unpredictable and complex. However, our results showed that procedures and the use of checklists and protocols promoted resilience by structuring parts of the work process. By creating procedures for uncomplicated events, one can control and reduce unwanted variability in these situations (Macrae, 2013; Hollnagel, Braithwaite and Wears, 2013b).

At both hospitals, statistics were made available for all employees, and this seemed to have a positive effect. Informants gained a sense of "ownership" of indicators and they were used both to monitor their own and other hospitals' performance, and as a source of learning and improvement. Results of comparisons were used to assess whether the maternity ward could learn something from other hospitals with better statistics than themselves. We argue that the implications for others will be to focus on information access and create a collective interest for the everyday results on the wards. The development and use of local and national indicator data as a source of information to monitor and follow up everyday practice is probably unexploited in health care.

In our study, practical training on site appeared as a key resilience mechanism to foster learning. Midwives took the initiative to implement exercises when time permitted. Practical exercises were also given priority in internal and external courses. The maternity wards used these courses to improve the ability of staff to respond to both expected and unexpected situations. Our results confirm those of other studies, showing a relationship between simulation and ability to provide safe patient care (Siassakos et al., 2013; Aase and Wiig, 2010). Simulation has several purposes. First and most obvious, it increases the employees' knowledge and improves their technical skills. In addition, simulation improved the ability of interprofessional collaboration that staff called for in our study (Vincent, 2010; Siassakos et al., 2013; Ennen and Satin, 2014).

The maternity services at the two hospitals differed in the quality of interprofessional teamwork. Informants at hospitals 1 described a good interprofessional teamwork and pointed out the advantages of being a small women's clinic with the benefits of close cooperation, good knowledge of colleagues at all levels and a collective sense of responsibility. These are factors that can contribute to a good collaboration climate and cohesion (Hollnagel et al. 2013b; Vincent, 2010), which our study found as important elements of promoting resilience.

Conclusion

Several similar mechanisms were involved in shaping resilient maternity services across hospital contexts, although the university hospital had a larger repertoire of mechanisms that could be categorised under each resilience cornerstone. Anticipation mechanisms were related to competence, team composition, organisation, procedures and routines. Monitoring mechanisms involved proactive and reactive indicators to anticipate and respond by evaluating and adapting their clinical practice. Learning mechanisms involved simulation, theoretical courses, internal professional updates and training scenarios to improve professional, technical and non-technical skills in daily practice. Response mechanisms were related to resource availability, flexibility, collegial and professional support, inter-professional collaboration, organisational learning and IT systems.

The main differences between the two maternity services were related to mechanism repertoire under each cornerstone, and the degree of openness and collaboration between professional groups. The rural maternity ward fostered a better learning environment than did the city-based university hospital.

Resilient maternity services depend on multiple mechanisms that contribute to the ability of anticipation, monitoring, learning and response, and most importantly the dependencies between them. The particular context of each micro system under consideration here is vital in understanding resilience. Therefore, we argue that a process of organisational diagnosis could be a starting point for practitioners (e.g., senior managers, ward managers, micro-system staff) with interest in promoting resilience in their organisation. One way forward is to use the four cornerstones of resilience (Hollnagel, 2014a), or a resilience model (Lundberg and Johansson, 2015), and identify the content or mechanism repertoire of each cornerstone, and evaluate the quality and variability in these. By identifying functions and possible weaknesses in their performance, improvement measures can be developed and implemented to promote resilience. This is a possible way forward in order to operationalise the resilience theory and transfer it into possible interventions in different contextual settings. This is currently lacking in the resilience research (Righi et al., 2015).

Finally, we recommend further studies of why different clinical services (e.g., maternity, cancer care) succeed, in different contextual settings (large, small, rural and urban). By carefully mapping and incorporating context sensitivity in study design, future studies can enhance our understanding of context as a key element in producing different repertoires of resilience mechanisms, which can explain when and why things go right.

Chapter 9

Team Resilience

Implementing resilient health care at Middlemore ICU

Carl Horsley, Catherine Hocking, Kylie Julian, Pamela Culverwell and Helmer Zijdel

Introduction

Resilient health care (RHC) represents the application of resilience engineering principles to the health care system. In this relatively new field, most of the writing about RHC has been by safety scientists and researchers but it is now timely to look at how clinicians might apply this new way of thinking in everyday practice. This chapter discusses how RHC principles have been incorporated into a general intensive care unit (ICU) and the effects this new approach has on the delivery of patient care.

Background

The Critical Care Complex (CCC) at Middlemore Hospital provides intensive and high dependency care in South Auckland, New Zealand with over 1,200 admissions per year including 200 children. It is the national ICU for major burns as well as one of two national centres that care for patients with acute spinal cord injuries. It has 70 per cent acute admissions and has outreach and emergency response roles throughout the hospital. The patients are from diverse multicultural backgrounds as are the 150 staff. It is a closed unit, which means admission decisions are made by intensive care senior doctors (intensivists) and management is carried out by the CCC in conjunction with other services as needed.

Quality improvement has been a strong focus for the CCC, which has a dedicated quality co-ordinator who has strong links to the clinical and education teams. The unit has led many initiatives including the implementation of bundles of care for preventing central line infections (Seddon et al., 2014), ventilator associated pneumonia, pressure areas and delirium.

Additionally, there is an education team comprising medical and nursing staff who have worked together since 2011 to provide an interdisciplinary simulation programme for team training and human factors education. The programme is linked to both nursing and medical curricula and evolved from single profession simulations performed in the local simulation centre to monthly full team simulations carried out on shared study days within the CCC using normal bed spaces, equipment and medication. This development was supported by local management who saw it as a cost of providing high-quality care.

While the quality and education programmes had led to improvements in the unit, there was a realisation that the existing approaches had almost reached their limits. The approaches to safety based on compliance, incident reporting and retrospective review did not seem to be reducing the incidence of events and the simulation programme had not produced consistent teamwork behaviours on the clinical floor.

As part of exploring potential ways of progressing these issues, we read about Hollnagel's work on resilient health care (Hollnagel, Braithwaite and Wears, 2013c) and wished to explore whether this could inform our own practices of care at CCC. What follows are the methods of this process, including the development and implementation of a framework, some findings from interviews with staff, and also some lessons learned about implementing cultural change in clinical settings.

Developing a model

Hollnagel defines resilient health care (RHC) as 'the ability of the health care system (a clinic, a ward, a hospital, a county) to adjust its functioning prior to, during or following events (changes, disturbances, opportunities) and thereby sustain required operations under both expected and unexpected conditions' (Hollnagel et al., 2013c; Wears, Hollnagel and Braithwaite, 2015a). It is important to understand that it is the ability to adapt over multiple timescales that marks the concept of resilience as different from concepts of robustness or rebound, in which temporary stressors on the system (i.e., patient admissions, acute events, disasters) must be absorbed without overt failure (Croskerry, 2009). Accordingly, our aim was to increase the capacity for resilience in the CCC.

The approach we took was informed by an understanding of intensive care as a complex adaptive system (Braithwaite et al., 2013). It operates at the junction of urgency and complexity (Cook, Woods and Miller, 1998) where multiple technologies and a diverse team of people work to care for critically unwell patients. It is also important to understand that there is no 'usual care' in intensive care. While there are consistent approaches, the variety of presentations and the individual responses to therapy mean that intensive care is always somewhat individualised, tailored to the needs of the patient and their differing responses to treatment. These system and patient factors mean that safety in the ICU varies constantly in response to the demands and constraints placed upon the system.

RHC is grounded in this more complex view of health care and the implications it has for how patient harm occurs (Hollnagel, 2013b) and the ways in which we can enhance the resilience of the system by increasing the capacity to monitor, respond, anticipate and learn (Hollnagel, Braithwaite and Wears, 2013b).

In this changeable environment, complex problems arise that may exceed the capabilities of any one individual and require multiple viewpoints to deal with them (Ashby, 1956). Communication is therefore not just about the transmission of information but how the ICU team constructs and constantly updates a shared understanding of the situation that enables all the resources of the ICU team to be utilised (Eisenberg, 2008).

The aim of the programme therefore was to develop adaptable teams and to increase the resilience capacity of the CCC to monitor, respond, anticipate and learn.

The Team Resilience Framework

The principles above seemed relevant to almost all aspects of our work within the CCC and we therefore developed a framework based on the work of Mark Meyer (Kaiser Permanente, San Diego), which we could use to embed these across our work. The Team Resilience Framework (the Framework) was designed to be scalable across all aspects of team function in the CCC from meetings, to crises, to the ward round and even for patient-centred care. It describes the elements of a functional team and how this forms the foundation for being able to adapt when conditions change. It is not about building a rigid team structure but rather it provides a shared understanding of the usual way we do things in order to understand when it might need to change.

It brings together the concepts discussed above with some of the familiar tools used in team and human factors training but focuses on how the tools can be varied for use in different contexts. For example, a Situation-Background-Assessment-Recommendation (SBAR) handover will be different in a cardiac arrest situation from how it will be on an ICU ward round. Likewise, pre-briefing discussions can vary markedly depending on urgency, complexity, team size and composition. There is no specific way that will be appropriate for every situation and the form, the amount of information and the words spoken will be different. It is this context sensitivity that we discuss and model with our teams, not just the tools in isolation. It has structured variability: the elements are consistent but the content is context sensitive.

It should be noted that there is a degree of dependency between the Framework items. For example, without understanding the situation, the team cannot accurately predict the resources and roles required. Likewise, if the plan is not explicit then the team cannot identify that it is not working. The Framework should therefore be viewed, to some degree, as a sequence.

The Framework

1 'DOES everyone know what we are trying to do?'

A shared understanding of the current situation is the key starting point for any team. When the whole team knows the problem and plan, it is better able to prioritise and anticipate issues, and the resources of the whole team can be utilised, rather than just those of the team leader.

A shared understanding can be built using the following sequence:

- SBAR: Setting the team in the right direction – The SBAR (Situation – Background – Assessment – Recommendation) tool is useful to describe the situation and clarify the issues. It is highly scalable and is used for routine handovers, ward rounds, referrals and clinical emergencies. It ensures that everyone understands what has happened so far rather than making assumptions about what has happened.
- Prebrief: Making explicit the situation and the priorities – Following a brief assessment, it is vital that the team leader makes the problem explicit for the whole team and clarifies the priorities. Making this explicit reduces the assumptions people make and enables people to anticipate the likely tasks and resources required. It also provides an opportunity for the whole team to add any missed information or corrections.
- Recap: Keeping the team on track – A recap is a brief summary of the situation and the current priorities. It should be used to ensure the team remains on track. It may be useful when: there is a significant change or if the situation is 'not going to plan', significant numbers of new team members have arrived or as requested by the team to clarify the current situation and priorities.

2 DOES everyone know who is doing what?

Once the situation and priorities are clear, the team can then focus on who covers the required roles. Roles can either be allocated or self-nominated, the more important thing being that everyone in the team knows who is doing what and what jobs have been assigned.

- Leadership – It is difficult for the team to work effectively when there are multiple competing priorities. The key roles of the team leader are to establish the team climate, make the priorities clear and provide the strategic overview of the situation. It is important to understand that leadership is not about seniority, it is about the role for the team. This separation enables senior staff and different specialities to be part of effective teams without precipitating leadership conflicts. Handing over the role of strategic overview should be made explicit to the whole team.

- Different leadership styles are required for different situations. While a more directive approach may be required in a relatively simple crisis, distributive leadership is more effective in bringing the entire resources of the team to bear on more complex problems. The team leader always has authority but does not need to be authoritarian.
- Active followership – Active followership means that there is an expectation that the whole team contributes to solving the problem. Good solutions can be used regardless of who has them and expertise can be used even when staff are not in the leadership role. There is a shared responsibility for solving the problem and ensuring the team functions well.

3 'ARE we clear in our communication with each other?'

Being explicit and efficient in communication reduces the mental workload for team members. The team needs to be clear about WHO is being tasked, WHAT is being asked for, and letting the team leader know WHEN it has been completed.

- Direct communication – Using names and making eye contact ensures it is clear who is being assigned the task.
- Reading back – We need to make sure that communication is accurate at all times, even during highly stressful events like a resuscitation. Reading back helps to ensure the message sent and the message received are the same. It is particularly important when verbal orders are given, as miscommunication can lead to wrong drug, wrong dose or wrong unit errors.
- Closing the loop – In complex and dynamic situations it is extremely difficult to keep track of multiple tasks and people. Closed loop communication provides feedback to the requestor that a task is completed. This reduces the cognitive load on the team leader by decreasing the need to keep track of all tasks and team members.

4 'HOW do we achieve our goals: even if things change?'

- Anticipate – Once there is a shared understanding of the situation, the whole team can anticipate the likely course of events including how things should go, as well as possible ways things could fail. Being explicit about expectations makes it easier to identify when things are not going to plan, even if it is in a way we hadn't considered.
- Monitor – All members of the team need to be clear about what to monitor in order to identify if the plan is succeeding or not. By knowing what to look for, we can respond more quickly to changes when they occur.

- Respond to what is happening – We need to be able to respond flexibly to events that are happening, even unexpected ones. It is important that changes in the state of events be made explicit to the whole team and new priorities established.
- Learning from all events – It is vital to set aside time for the team to consider any insights gained from the situation. Previously we only reflected on our practice when something 'went wrong' but it is just as important to learn why things went 'right' as often the team has had to work around problems to create the good outcome. Understanding what led to the good outcome despite the difficulties helps to make success reproducible. Debriefing of clinical events is an example of building this reflection into clinical practice.

5 'HOW do we speak up if we have concerns?'

If we have a clear understanding of the situation and can see things that concern us or don't fit, we need to be able to speak up to ensure the safety of the patient. The aim is to provide an escalating response that gives team members the ability to raise concerns about safety and still maintain good working relationships within the team, both now and in the future.

- Enquire to ensure that there is a shared understanding of the situation.
- Advocate for a different approach; challenge; ask a curious question.
- Assert the need to do things differently; call for help; take over.

6 HOW do we foster a safe team environment? (psychosocial safety)

If people don't feel safe within the team or system, then they won't speak up or contribute. Team leaders have a vital role in ensuring an open and productive team climate that enables all team members to contribute to solving the problem.

- Reducing hierarchies – Hierarchies between and within groups act as potential barriers to effective teamwork and communication. The team leader can reduce these by using briefings to promote a positive team concept i.e., we are all working together towards a common goal. The team leader can also promote speaking up by inviting questioning or cross checking e.g., 'if you see anything that doesn't seem right then please let me know'.
- Valuing speaking up – If a team member has had a negative response from someone previously, then they are less likely to speak up, even if there is a patient safety issue. Acknowledging and valuing suggestions, even if they are not acted upon, means that team members are more likely to speak up in future.

- Focus on learning – How does the team or organisation respond to errors? Do we focus on blame (e.g., do we ask 'how could they have done this?') or on learning (e.g., do we ask 'why was it that seemed the right thing to do at the time?'). Our responses will affect both how people speak up and also how they view errors by themselves and others in the future.

Implementation

The Framework was introduced in January 2014 to the group of CCC staff who had previously been involved in the simulation programme. The group, now known as Team Resilience, included intensivists, charge nurses, senior nurses and the education team.

The implementation of the Framework was an iterative, adaptive process as it was unclear what effects if any would be seen and how it would impact on current initiatives. The most significant progress occurred when the Team Resilience members applied the Framework to how they organised the simulation days. This allowed them to practise the concepts and reflect on ways the Framework could be built in to everyday work.

Spreading understanding

Nursing and junior medical staff were introduced to RHC principles such as Safety-I and Safety-II, Work-as-Imagined (WAI) vs Work-as-Done (WAD) and concepts of resilience through presentations, posters and informal discussions. The aim was to provide some background understanding to the Framework. The concepts were also presented to members of the safety, quality improvement, infection control and emergency departments to give some of the departments that interact with the ICU an understanding of what we were trying to implement.

Showing how the Framework worked

The interdisciplinary simulation days were reoriented to show how the Framework might work in practice. The learning objectives and Framework are made explicit before the simulations. The scenarios still use acute events as these are emotionally engaging and demonstrate the team and system under stress.

The first scenario on a simulation day is designed to demonstrate how to create functional teams (shared mental models, clear roles, clear communication) while the second scenario shows how an effective team can adapt to meet unanticipated problems using resilience principles (anticipate, monitor, respond, learn).

The debriefing discussion has changed to look at such concepts as:

- Shared responsibility: e.g., asking for a plan, roles and mutual support.
- Context sensitivity: How would the tools vary in different situations? How would leadership and followership change?
- Understanding usual work: why things work well, not just how they go wrong.
- Workarounds and suggestions to make work easier.

The debrief also discusses any problems with processes or equipment and any other issues that make usual work more difficult.

Modelling and practice

While the simulations were useful to show how the change in behaviours affected team function, staff needed to have the opportunity to practise and integrate the principles. Rather than just hope for the training to change practice, the members of Team Resilience also consciously modelled the Framework in clinical practice such as in the response to acute events and on the ward round.

Acute events

The timely management of critically unwell, poorly differentiated patients is a key capability of the ICU team. The aim was to reinforce the simulation teaching with the same sequence of team construction and focus on anticipatory briefing. The elements from the Framework are consistent although the discussions are always unique and tailored to the problems of the patient.

An intubation checklist from another hospital had previously been introduced but had not gained wide usage. The education team developed a new checklist based on the Framework that was designed to be a cognitive aid that, in addition to the usual equipment checks, prompted staff to discuss:

- Anticipate: What should happen – what is the intubation plan for this patient?
- What could happen – what particular hazards are there?
- Monitor: What will we monitor to signal the need to change our actions?
- Response: What is our plan to respond to the hazards we see?

The aim was not to describe what a generic intubation should look like but rather was about building a shared understanding of the particular plan and risks for this patient.

A further example was changing the expectation about debriefing following an event. Previously, debriefings only occurred in the setting of failure;

either a poor outcome or some problem with the team processes. There is now a very brief discussion held in the bedspace immediately after all major events (resuscitation, intubation, major procedures etc.) guided by the following questions (Conklin, 2012):

- What happened as we expected?
- What surprises were there?
- What hazards did we identify and what did we miss?
- What issues came up that we had to work around?

These questions help to identify unexpected problems the team had to work around but also to learn how the team dealt with them to ensure success. Any equipment or system issues are forwarded to the quality coordinator as well as suggestions from the team regarding how they could be resolved. These discussions have been useful in demonstrating that successful outcomes occur not because there were no problems but rather because the team was able to adapt and find ways around the problems.

Ward round

The CCC ward round is a key time when the team as a whole comes together to make decisions about patient care. It is often assumed that by coming together the team automatically develops a shared understanding of patient care but this is not necessarily the case (Reader et al., 2011). This may result in an inability of team members to contribute effectively and lead to uncoordinated care. This had previously manifested in the CCC as a focus on task completion without awareness of the overall goals and also inconsistent messages to families due to the team having differing understandings of the patient situation.

The ward round at CCC had evolved over the few years prior to the Framework, with separation of the medical handover from the multidisciplinary review at the bedside. We introduced the Framework in both aspects of the ward round as discussed below. The changes have not increased the overall length of the ward round.

Medical handover

The medical handover is a short daily meeting of intensivists, registrars, charge nurses, pharmacists, dieticians and the unit manager. This occurs away from the bedside and first focuses on developing an understanding of the situation for the unit as a whole. It includes discussions on demands (current bed state, patients under review or awaiting a bed, bookings) and constraints (staffing issues, hospital bed state, potential discharges). This discussion has evolved over time and has helped us to be able to anticipate the likely problems at the unit level over the day. Additionally, it means our decisions are made with an awareness of the overall situation in the unit and the hospital as a whole.

Second, the medical handover of patients follows an SBAR structure that varies with the complexity of the patient. Other team members can correct or add to it as required and there is discussion to clarify relevant information. The aim is to develop a coherent narrative for the team as a whole. There was an increased focus on being explicit about the likely projected course for the next few days, potential concerns and the decisions needing to be made by the incoming team. The aim was that a fuller understanding of the situation allows the incoming intensivist to crosscheck as well as helping to develop consistent plans across multiple days in highly complex patients.

Bedside multidisciplinary team review

The team review occurs at the patient bedside and brings together the medical, nursing and allied health team members as well as patients and families where possible.

The round starts with a brief summary of the understanding of the current situation and the issues to be addressed based on the handover. This is not meant as a definitive statement but rather as an opportunity to make sure no new problems have occurred or been missed by the handover. This statement also has a team building role: it brings the team together around a shared problem and includes an invitation to correct or add to that understanding e.g., 'Is that correct? Have I missed anything or has anything changed?', and is used as a way of creating a climate where contribution is expected (Nembhard and Edmondson, 2006).

Intensive care requires the whole team to be focused on what they can do to keep the patient progressing towards recovery. It also means rapidly responding to any problems that could set back their recovery. RHC principles are integrated into the ward round by thinking about and making explicit the following:

- Anticipation: What should happen – what are our expectations for this patient?
- What could happen – what hazards do we see?
- Monitoring: How would we know if things were on track or changing?
- Response: How would we respond to the hazards we see?
- Learning: What went as expected and what did we miss?
- What did we have to work around and what made it safe?

While we try to identify any major hazards, there is always the potential for unanticipated problems. Being explicit about the expected course enables the team to be clear when we are seeing something unexpected. The hope is that this leads to earlier identification of problems and a quicker response.

Previously, the intensivists often had an internal concept of the likely expected patient course and likely hazards but did not consistently share these with the team. However, this risked leading to a situation where the other members of the team made assumptions that were not aligned with the intensivists' priorities. It also meant that at time of crisis the intensivists were potentially task overloaded. By making this thinking explicit and pre-planning responses, the team can better prepare and it also reinforces the concepts of shared team responsibility and anticipation, even if no untoward event occurs.

The bedside round finishes with a recap of the plan, the expected course and the potential problems. Again, this is a time for the staff to remind the intensivists about things they have missed or clarify questions about the plan. It is a plan that is constructed by the team and, whenever possible, by the patient as well.

Evaluation

Method rationale

When the project was commenced it was unclear what changes, if any, the introduction of RHC concepts and the Framework would have on everyday clinical work within the CCC. Unlike usual quality improvement interventions, this was not a single intervention to be measured in a controlled way but rather involved introducing and embedding principles that could impact on many aspects of our work, over a wide timescale and potentially in unpredictable ways.

As such, this initial report represents our attempts to get viewpoints from a variety of staff within the CCC on changes seen since the implementation of the framework. The data sources were primarily qualitative involving interviews and in-practice observations. A qualitative, evaluative design was used to assess staff perceptions of the changes they had seen in their everyday practice over the 2 years since the Framework was implemented.

The in-practice observations were made during and since the implementation phase of the Framework by the authors of this chapter, who all worked within the CCC.

Data sources

A qualitative, evaluative design was used to assess staff perceptions of the changes they had seen in their everyday practice over the 2 years since the Framework was implemented. Purposeful sampling recruited 24 nurses and four intensivists who had worked in the CCC for the entire 2 years since the introduction of the Framework. This excluded all the junior doctors, who work in the unit on 3–6 month rotations. CCC staff involved in the rollout or development of the Framework were also excluded from the interviews. Ethics approval was sought and waived by the local ethics committee, as they deemed it a quality improvement project not requiring their approval.

Of the nursing staff, eight were junior critical care nurses, twelve were senior nurses who had completed a recognised critical care education programme and four were charge nurses. All groups were highly experienced with an average time in the CCC of 6.5 years for nurses and 12 years for charges nurses and intensivists.

The wider in-practice observations occurred as new developments were seen within the CCC, which were judged by the authors to be informed or influenced by the Framework or wider RHC principles but were not part of the original implementation.

Data collection – interviews

In December 2015 the CCC quality coordinator (2nd author) interviewed the 28 medical and nursing staff. The interviews were semi-structured and asked:

- What changes, if any, have you noticed in the way the team works together?
- What differences, if any, have you seen in the way the team works when they are dealing with situations such as intubations, cardiac arrest or other types of emergencies?
- Thinking about the ward round, can you identify any specific behaviours that have changed?
- How do you feel these changes you have identified have impacted on overall patient care in our unit?

Interviews were on average 7 minutes long (range 3–25 minutes). They were recorded electronically and transcribed verbatim by the CCC unit secretary and were subsequently checked for accuracy by the interviewer. Staff involved in the rollout of the Framework were excluded.

Data analysis – interviews

The interviews were analysed using inductive thematic analysis (Braun and Clarke, 2006), which involves highlighting interesting phrases or features within the interviews. Recurrent terms or ideas were organised into groups and coded manually. An iterative approach was taken in which the data and categories were reviewed and collapsed until larger, overarching themes that remained coherent with the original data could be identified.

Results of the thematic analysis

Thematic analysis identified three key themes: improved team organisation, improved psychosocial safety, and proactive safety behaviours. Each of these themes and the reported impacts on patient care will now be discussed.

Improved team organisation

Almost all interviewees reported that there was a more planned approach both in the response to events and on the daily ward round. There was a shared expectation of what should happen and how the team should organise itself.

> *We are less dependent on individual experience now that people work together better, and have a bit more of an expectation about how things will go, that we will talk about what we are expecting. (Nurse, CCC)*

Most respondents reported a greater sense of the team having shared goals and 'being on the same page' as well as having greater clarity about the likely roles that would be required.

The doctors reported that nursing staff were increasingly self organising and would ask for the information they required if it was not forthcoming.

> *All that leads straight away into our usual thing of 'what's our plan A and what's our plan B and what's our bailout?' So that all happens pretty much automatically now and as I say, if you do not voluntarily do it, people will ask because they will want to know. So they will ask 'what will I be doing, who will be that person?' In that sense, it sort of starts to become a culture thing that people expect, then they ask for it if you do not automatically do it. (Doctor, CCC).*

Most respondents reported a greater sense of the team having shared goals and 'being on the same page' as well as having greater clarity about the likely roles that would be required.

Many responses mentioned the value of the checklist as an improvement in team organising for intubations. There were differing opinions regarding how it should be used, with some stating it was useful to be able to follow the steps while others thought it was a way for the team to ensure nothing had been missed once they had prepared everything.

> *It probably took 40 seconds to completely cover off everything on that checklist and that was because things were already being prepared. People*

were already realising these roles needed to be covered and that we would anticipate certain things happening. That was happening without the most experienced team on. (Doctor, CCC)

Improved psychosocial safety

Both nurses and doctors reported a sense of inclusiveness and of being part of a cohesive team beyond professional groupings.

The whole culture has changed and I think it has become a really focused group effort department with everyone looking out for each other and working for each other and with each other. (Nurse, CCC)

Nursing staff felt that the medical staff were more approachable and encouraged the nurses to ask questions or share their viewpoint on the ward rounds. While some felt that being asked questions had previously seemed intimidating, they now understood it more as a group discussion. Many respondents reported feeling more comfortable to speak up about their concerns and to contribute to discussions.

I have noticed the team environment makes it not so scary to say to the specialist 'I don't know', 'I don't know how to do that' or 'I think he is not ready' or 'we have been struggling to get him out of bed'. You know it seems like it is more of an open discussion as opposed to 'you will do this, this and this today'. (Nurse, CCC)

When they did contribute, nursing staff felt they were being listened to and as a result felt valued. A significant number of respondents, both medical and nursing, reported that staff in the CCC were happier and less stressed.

Proactive safety behaviours

Staff reported a greater use of pre-emptive plans to deal with potential events. There were frequent descriptions of multiple sequential plans for failure (Plan A, Plan B, Plan C) as part of briefings during incidents. Staff noticed a change in the discussions regarding identifying hazards and planning a response.

If you have got someone with a difficult airway, and they lost their airway, what would you do? What scenarios? It is quite good just to have that in the back of your mind. Nine times out of ten it does not happen but I think it is good to be prepared and just to have people thinking about it and then if it does happen you do know what you're going to do, you have a plan in your head. (Nurse CCC)

The medical staff reported that they had an increased focus on sharing what their expectations were, including the possible ways in which the patient might deteriorate.

> *So, for me the difference is sharing that worst case scenario saying 'I don't think that is going to happen', 'this would be the worst thing', 'what I think is going to happen here is', 'it would be terrific if this happened' so at least people can see where we are in the road. Are we on this kerb or that kerb? Where are we going? I think that has allowed people to relax a little bit and focus on good care rather than everything being a surprise. (Doctor CCC)*

Effects on patient care

Almost all respondents felt that the observed changes in team behaviour and communication led to better patient care although some respondents expressed uncertainty as to how the perceived improvements in care could be measured.

> *I think anything that means you have a more cohesive team has got to be better for the patient and that's the bottom line. If you have a team that works together more, you sort of get an idea of what you are aiming for, you are going to get a team effort and you will get more consistent care. You will get differences of opinion but you will get more consistent care. (Nurse, CCC)*

> *I think we are more prepared in that when things go well, it's not just luck. We are more prepared and so when things are not going well you are able to step in sooner. . . .. You may not be able to stop [the deterioration] in time rarely but for some people those little tweaks can make the difference rather than 'oh no, they have fallen over completely and now we have to try and get them back up'. (Doctor, CCC)*

In-practice observations

In addition to the changes described in the interviews, the introduction of RHC into the unit had other observed effects, including the following examples.

Improvement initiatives following routine debriefing

The routine debriefs following an event identified several areas that made routine work more difficult and increased the likelihood of harm. These included such issues as poorly laid out drug refrigerators and equipment failures as well as process issues both within the CCC and between departments. Staff would frequently suggest ways in which the process or issue could be improved to make their work easier.

An example of staff-suggested improvements related to the handover discussions when receiving patients from other parts of the hospital. These were often chaotic, especially with critically unwell acute admissions, with a lot of activity that was not always coordinated. There were often multiple simultaneous handovers given (e.g., nurse to nurse and doctor to doctor) yet the receiving team often required more information after the handing over team had left. While checklists and highly standardised approaches have been used successfully in other hospitals, these are often based on more stable and standardised post-operative patients. By contrast, the high acute load and large variety of clinical problems meant a highly standardised structure would be unlikely to work in the CCC.

The handover project was a spontaneous improvement process initiated by staff. There were only two requirements: 1) make the patient safe (e.g., transfer to a ventilator) and 2) everyone stop and listen to the handover. Team Resilience asked the staff to try different ways and work out what made the most sense to them. The handover process has ended up being faster, clearer and requires less repeated information. It has now been used successfully for the full range of patients including cardiac arrests and emergency intubations. We have now seen widespread adoption of this new approach despite no formal rollout or intervention.

Impact of Work-as-Imagined (WAI) vs Work-as-Done (WAD) concepts

A further observed effect was in the work with the Infection Control team regarding hand washing. Hand hygiene is considered one of the markers of quality of care but despite this our compliance rate was relatively resistant to change from the level of 75 per cent. When trying to work out why, it is important to understand that there may be up to 150 hand hygiene moments in a 12-hour ICU nursing shift (Goodliffe et al., 2014). This represents up to 75 minutes of handwashing for a nurse who is also trying to care for a critically unwell patient. The need to navigate the competing demands of caring for a critically unwell patient and still be compliant with handwashing protocols represents an example of the WAD by the clinical staff and the gap between this and WAI by the infection control team.

Previously the gap was seen as a failure of compliance but the Infection Control team has now approached this from an RHC perspective to see it as an issue of balancing goals and making necessary trade-offs. The team came and spent time with CCC staff to understand how normal work was done. The focus was on understanding everyday clinical work and finding ways of building hand hygiene into staff workflows. It changed the way the Infection Control team taught hand hygiene and allowed staff to contribute to solving the issue. Staff are still audited against the WHO 5 Moments (Pittet et al., 2009) and this has shown an improvement to rates approaching 90 per cent.

The final change was in how safety incidents are investigated in the CCC. The quality coordinator adopted the concepts of WAI vs WAD and focused on understanding how the staff members' actions made sense to them at the time. This led to a deeper understanding of how normal work was done and what was different about a particular event. For example, this approach was used in response to medication safety incidents in the unit and provided more understanding of the demands faced by staff and how they normally make medication dispensing safe. This was in contrast to the previous approaches focusing on deviations from protocols with the subsequent remedies of education and reinforcing compliance.

Data interpretation

The interview themes and in-participant observations must be interpreted in an attempt to develop a coherent understanding of how the changes seen contribute to increasing resilience in the CCC. This interpretation is informed by the model of resilience developed by Hollnagel (Hollnagel et al., 2013b), which examines the ability to respond, to monitor, to learn and to anticipate.

Discussion

The role of teamwork in resilience

There have been positive changes in team behaviour over the last 2 years. These are over and above the changes achieved in the previous 3 years with a mature simulation programme in place. The interviewees perceived improvements in team organising, psychosocial climate and an anticipatory focus. It is particularly striking that nursing and medical staff alike gave consistent positive feedback given that nursing staff often report a poorer view of team functioning and communication than medical staff (Thomas et al., 2003).

For the CCC, team training and human factors teaching had the greatest impact when the focus was expanded from in-centre and *in-situ* simulation to finding ways to teach, model and weave the RHC principles into all aspects of daily work, including the simulation programme, the way we organise care and our approaches to quality improvement.

The importance of effective teamwork in health care has been extensively studied both from a deficit viewpoint where poor communication and teamwork have resulted in adverse events as well as where good teamwork and communication have been associated with improved team function, clarity around daily goals, staff wellbeing and improved patient care (Manser, 2009).

The resilience engineering perspective helps contextualise these findings by considering how team function affects the capacity for resilience. Viewed through this lens, the changes implemented in the CCC increased the resilience capacity by:

Increased capacity to anticipate

- Developing teams that can engage multiple viewpoints to build improved situational awareness.
- Handover discussions of anticipated constraints/demands.
- Ward round discussions of expected course and potential hazards.
- Pre-briefings for acute events describing plans and potential hazards.

Increased capacity to monitor

- Ensuring all team members know the plan so more people can monitor for changes.
- Improving the ability of any team member to raise concerns effectively.
- Being explicit about the expected path to enable recognition when things change.
- Being explicit about the thresholds for action in an acute event.

Increased capacity to respond

- Improving team organisation to allow rapid response.
- Having plans for response already discussed on ward round and during acute events.
- Creating a team structure that allows all the team to help solve the problem.

Increased capacity to learn

- Debriefing of clinical events, even when they went well.
- A focus on incidents as an opportunity to learn about usual work and conditions.
- An improved psychosocial climate enabling solutions to come from anywhere.
- Reducing the gap between WAI and WAD to enable learning from frontline staff.

It can be seen that many of the changes made that impact on resilience capacity come from improvements in team functioning and the enhanced ability to identify and respond to problems. This is consistent with observations of resilience in surgical teams whereby debriefing successful operations identified latent threats that had been dealt with by the team (Catchpole et al., 2007).

One of the interesting aspects reported was the way in which the team became self-organising. The most striking manifestation of this was that the intensivists who had not initially been involved in the original training noticed they were being asked for pre-briefing, anticipatory plans and debriefs. The concept of active followership and mutual support meant that the followers were requesting the leadership behaviours they needed for the team to function. This is in contrast to previous studies that have focused predominantly on the leadership behaviours as the key determinant of team function (Manser, 2009).

The need for the 'capacity for learning'

There has been a growing acknowledgement of the need for adaptable teams that can monitor, respond and anticipate, particularly in highly dynamic areas such as intensive care and emergency medicine (Vincent and Amalberti, 2015). The changes seen in the CCC, however, suggest that it is the 'capacity for learning' that transforms RHC from being about the ability to cope with dynamic situations to being able to adapt and improve. It was the shift to reflecting on how success was created that enabled the team culture to change so rapidly in the CCC. It was also the key element that led to clinician-generated quality improvement in the form of the handwashing and hand-over improvements, and highlights the capacity for innovation enabled by the RHC approach.

The checklist

In the interviews staff commented on the value of the intubation checklist although it was used in different ways depending on the situation. Checklists have been a favoured quality improvement intervention since the publication of several trials showing improved patient outcomes and improved team functioning (Haynes et al., 2009; Pronovost et al., 2006). However, the broader rollout of mandatory checklists has shown more inconsistent effects (Urbach et al., 2014) and Catchpole recently discussed how authentic checklist completion relies on pre-existing good communication and team functioning (Catchpole and Russ, 2015). If the underlying sociological and cultural issues within the team have not been addressed, then the checklist will have minimal impact. In this view, checklists do not create functional teams but, rather, functional teams use checklists in a way that has value to them. This is consistent with observations of how the CCC teams used the same checklist in different ways depending on the context. It appears that CCC staff preferred to have a usual approach that could be adapted or even discarded if needed to match the situation.

Staff engagement

In the interviews, staff reported improved engagement and mood and this was confirmed by a recent survey of nursing staff engagement within the CCC. The sources of this increased engagement appear to stem from the improved psychosocial safety and mutual support within the team. Additionally, the RHC view of safety as an active creation represents an approach-orientated goal that has been associated with reduced burnout (Naidoo et al., 2012) and stands in contrast to the prevailing model of safety as a state maintained through compliance.

This is not a small matter with burnout being an endemic problem in ICU and staff wellbeing now being recognised as a pre-requisite for the delivery of safe, effective, patient-centred care (Bodenheimer and Sinsky, 2014). Enhancing group psychosocial safety is also associated with less emotional exhaustion (Welp, Meier and Manser, 2016), more engagement with quality improvement (Nembhard and Edmondson, 2006) and an increased likelihood of 'speaking up' (Szymczak, 2016). Any approaches that improve team function and psychosocial safety are therefore likely to have beneficial effects on multiple aspects of clinical care.

Implications for simulation

A final note is needed about the implications for simulation from this work. We had initially seen only small changes in clinical behaviour despite having a mature simulation programme in place. The changes to team behaviours we have subsequently seen are due to a shift from seeing simulation as an educational event used to change team behaviour, to seeing simulation as just a small part of an overall approach. Modelling and reinforcing what is learnt in simulations by interweaving it into all aspects of practice and grounding it in the realities of everyday clinical work is essential.

Increasing the capacity to anticipate

Limitations

This study is a retrospective examination of changes perceived by staff in the CCC. While these changes have been significant for the running of the unit and the wellbeing of the staff, the impact of these changes on patient care is harder to ascertain. A more comprehensive study introducing these ideas to an ICU would need to look at a variety of changes including measures of team performance, staff wellbeing and, most importantly, the outcomes and experience of care by patients and their families.

The way in which we embedded the RHC principles into daily work was constantly refined in response to the local context and the effects observed.

This local approach is one of the great strengths overall but means that the individual elements we used may only be partially relevant to other settings.

Conclusion

The concepts of RHC are a natural fit with clinical work and were readily accepted by staff in the CCC. The change in message from one of compliance and constraint to one of people, teams and learning has been transformative for the staff at Middlemore CCC. We believe the implementation of RHC principles has been successful because they:

- describe the system as experienced by clinicians i.e., dynamic, complex and unstable;
- value people and their ability to adapt; and
- engage people in thinking about how to make success more likely.

Our experience suggests that a shared team concept and psychosocial safety are key pre-requisites for developing resilience at the team and unit level. This has implications for our understanding of team function, how we design team training and our understanding of the sources of resilience in health care.

RHC represents a humanistic approach to dealing with the complexity of modern health care quite different from the current focus on production, managerialism and compliance. The implications of this approach go well beyond safety to include staff wellbeing, quality improvement, productivity, education and patient-centred care. As such, it may be a way of progressing many of the current issues facing health care that have been the most difficult to change. The aspirations of RHC should therefore be expanded beyond safety or even 'sustaining operations' to seeing the potential for this approach to advance health care towards the long held goals of safe, patient-centred care delivered by engaged staff.

RHC should be more ambitious, as the humanistic approach which is grounded in everyday work may be a key way of progressing many of the issues facing health care.

Chapter 10

Understanding normal work to improve quality of care and patient safety in a Spine Center

Jeanette Hounsgaard, Bente Thomsen, Ulla Nissen and Ida Bhanderi

Introduction

The following case study illustrates how the Resilience Health Care (RHC) principles can be turned into practice using the Functional Resonance Analysis Method (FRAM). The chapter highlights important general principles such as *breadth before depth* and *functional resonance*, and methodological 'tricks' that can be learned from the case.

The Spine Center of Southern Denmark is part of Lillebælt Hospital. The Medical Department of the Center is an outpatient clinic attending to 13,000 patient courses per year. Patients with back pain are examined and triaged to treatment in either (i) a local private setting, (ii) the local municipaty or (iii) hospital treatment, e.g., back surgery.

Hence the patient courses in the Medical Department are very short and call for a high degree of effective coordination both in-house and across health care sectors.

The clinical setting is multidisciplinary, represented by medical doctors, nurses, chiropractors, physiotherapists and secretaries.

A chain of three FRAMs, initiated by adverse events, were prepared in the period from August 2012 to August 2014. The first FRAM described how an adverse event, reported to the Danish Patient Safety Database, could happen due to the variability of everyday work, i.e., *normal work*. The aim of the second FRAM was to improve the effective and safe functioning of a specific activity, identified in the first FRAM. The analysis went into depth with this activity and its condition for functioning successfully. Lessons learned from FRAM initiated the third FRAM, which will be described in this case study.

Adverse events are managed systematically in the Quality and Patient Safety team by categorising all adverse events into general clinically relevant themes. In semiannual follow-up, patterns are discussed in order to select a patient safety theme for quality improvement. This approach to managing adverse events is reactive.

Table 10.1 Reported adverse events

	No. adverse events	Adverse events regarding patient identification	Adverse events regarding oversights
2012	48	14	N.a.
2013	48	6	14

In 2012 and 2013 a recurrent patient safety theme was recognised and addressed but persisted despite continuing PDSA cycles (see Table 10.1). The theme was: Patient identification and oversights that typically occur in the clinical management of referrals to other health care departments. This could involve events related to patient identification:

- Journal is written in the wrong patient file.
- MR results are scanned in to the wrong patient file.

Another example could be oversights:

- Referrals to MR scans and rehabilitation plans are not written as intended or are written but not sent.

PDSA cycles had addressed the attention towards the adverse events and the existing guidelines to be followed in regards to enhancing patient safety. This included specific national guidelines for patient identification, which are implemented and commonly known (Sundhedsstyrelsen, 2013; IKAS, 2012).

Example of linear problem solving of the patient safety problem

In trying to solve or reduce the number of adverse events, a workshop was held with attending colleagues from other hospitals in the Region of Southern Denmark. The purpose of the workshop was to address adverse events, sharing knowledge and advice. The assumption being that the adverse events occurred as a result of disruption, as the clinicians in a multidisciplinary setting often manage disruptions and interruptions in daily work. Hence the need to switch between journals. Advice and different solutions were discussed, such as:

- having two computer screens – one for viewing conference patients and one for ongoing work;
- using pockets cards;
- looking into IT-solutions.

Little was known about the patient safety problem within the organisation. The solutions seemed questionable as causes for the adverse events were based on too many assumptions and not related to the local context. They also seemed both costly and comprehensive and were furthermore based on experiences from hospital wards, which in their structure are quite different compared with an outpatient setting. Mindful of doubts about the exact content, a pocket cards system seemed complicated to develop and IT-solutions were considered too expensive and comprehensive and thus disregarded as an option.

Typically, the adverse events occurred in the administrative clinical follow-up when the patient was no longer present in the clinic. Furthermore, the processes involve relying on no fewer than six different administrative systems, which undoubtedly enhances the risk of adverse events. Present national guidelines do not address patient identification in the setting when the patient is no longer physically present, nor the number of, or complexity of, administrative systems.

The Quality and Patient Safety team found that a systemic approach was needed to investigate this patient safety problem in order to match and validate further quality improvement initiatives.

Rationale for the research design

FRAM was recognised as an appropriate analysis method to elucidate the variability in daily routines and understanding normal work. Daily adjustments are necessary in a clinical setting and are found to be *a black box* in which both appropriate and inappropriate daily processes should be investigated. In this perspective, the purpose of the FRAM case was to (i) generate new hypotheses to a persistent patient safety problem in the Medical Department and (ii) test a new analysis method (FRAM).

FRAM training courses were available in the Region of Southern Denmark. Moreover, FRAM had been used in previous case studies, hence the Quality and Patient Safety team initiated a FRAM project and aligned it with participation in the FRAM training course. Seeing that FRAM methodology should be learned from both a theoretical and practical perspective, the methodology was followed rigorously as described in the FRAM course handbook (Hollnagel, 2013a).

The focus was to elucidate variability in the daily setting. Therefore, the project design is qualitative and follows the principles from Grounded Theory by Kvale, in alignment with recommendations in FRAM methodology (Kvale, 2002). As opposed to earlier PDSAs this explorative methodological approach shifts focus from 'what goes wrong' towards understanding *normal work* and new insights into the organisation as a system. This was considered to be an important aspect to address throughout the project in order to ensure balance towards the aim of the project.

The FRAM project was organised in the Quality and Patient Safety team in the Medical Department. It consists of clinicians with different professional background with time allocated for Quality and Patient Safety work and is led by the lead nurse who is a member of the departmental management. The team holds (i) one medical doctor, (ii) one physiotherapist with a key function as risk manager, (iii) one secretary with a key function as risk manager, (iv) one chiropractor, (v) one nurse with a key function as hygiene coordinator and (vi) one ergotherapist. This setup enables relevant questions in regards to the project design to be addressed in both clinical and management perspectives, combined. Support at management board level was ensured by given mandate.

The FRAM project was performed by the two risk managers in the Quality and Patient Safety team. Clinicians' insights to normal work and routines are essential in FRAM methodology and therefore thoughts of addressing a complex and persistent patient safety problem in a local clinic setting were discussed. The reported adverse events on the patient safety theme to be investigated were reported from all professions in the multidisciplinary setting with diverse attentions respectively. How to address these and ensure participation and involvement was addressed in the setup of the project.

The head of the FRAM training course, the regional risk manager Jeanette Hounsgaard, supervised the project as the FRAM course proceeded, to ensure consistency and adherence to FRAM principles.

The Medical Department was introduced to FRAM and the project at a conference in order to address any questions among staff before initiating the data collection.

How data sources were selected

Data sources were chosen to account for (i) the multidisciplinary clinical setting, (ii) expertise in daily routines, (iii) a validated description of Work-as-Imagined.

The subjects for investigation: Patient identification and oversight in an administrative context were reported as adverse events by all professions represented in the Spine Center. Two physiotherapists, one chiropractor, one medical secretary and one registered nurse were interviewed respectively. The medical doctors were not able to participate due to understaffing. As the focus was to describe variability in the daily clinic setting, individual semi-structured interviews were chosen as data sources. These were found to be the best data source in regards to elucidate how work is actually done compared with Work-as-Imagined.

The informants were chosen for their high level of experience and knowledge in the clinic and not related to the adverse events. This was to ensure objectivity and to emphasise the learning perspective of the project.

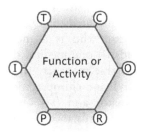

Figure 10.1 Visualisation of a function/activity with its six aspects.

A descriptive case was made in order to identify the functions of the study and illustrate the processes that were supposed to characterise how work is done but that rather represented Work-as-Imagined.

> *When the referral from primary care provider is received and pre-evaluated, the patient gets an appointment and is examined by a doctor, chiropractor or physiotherapist or allocated for MRI via a nurse. To make a plan for further examinations or discharge, a multidisciplinary conference is held; with or without the patient present. After the first visit the patient can be seen in the Spine Center for treatment over a short period, or for further examination. This might raise the need for other formal or informal conferences. Every visit or conference is concluded by writing journal, referrals etc. and the referrals are dispatched by the secretaries.*
> *Activity 1: To write a referral*
> *Activity 2: To dispatch a referral*

The risk managers collected qualitative data about the two activities, covering the six aspects of FRAM: input, output, control, resource, precondition and time (see Figure 10.1). Data included the variability of the output and the interactions/interdependencies with other functions/activities.

How data were collected

The two activities to be investigated were performed individually and the circumstances for doing so were variable. The aim was to look closely into the degree of details and variations in performing the activity, which called for a narrative approach (Kvale, 2002). The individually based semi-structured interview was chosen to ensure in-depth dialogue and new insights to daily variability. Each interview had a duration of 45 to 60 minutes. Five interviews were performed over a period of 4 weeks.

Table 10.2 Interview guide by Ulla Nissen and Pernille K. Clausen

Aspect	Interview guide
Input	Will you please describe the activity? When do you typically start the activity?
Output	What is the output of the activity?
Control	How do you identify your patient? What do you do in case of insecurity about the procedures? How predictable is the relationship between work conditions and the situation? Do you use memorising lists? Do you often change your work procedures? Is there a superior way to work?
Resource	Where do you do the activity? What kind of information do you need and is it available? Are the persons you need always present? What kind of skills do you need? What do you do if something unexpected happens?
Precondition	What should be in place before you start the activity? How stable are the work conditions?
Time	How do you deal with lack of time? Are there issues you often have to deal with under normal circumstances?

The interviews were prepared at the workplace of the informant. The interviewer briefed the informant about the purpose of the interview – *to identify quality improvement measures by describing how the job is done on an everyday basis (work-as-done), including necessary adjustment to the actual conditions to perform successfully.* In addition, the informant was briefed about the use of the final report and the following steps.

The interviews were prepared and managed by the use of a specific interview guide, covering the six aspects (Hollnagel, 2013a). Of the six aspects, *Control, Time* and *Resources* were emphasised in the interview guide. The questions used to elaborate the six aspects are shown in Table 10.2.

The two risk managers who performed the interviews also have clinical functions in the regional Spine Center and thus have in-depth knowledge of daily routines. Precautions for complicity were accounted for in each interview by appointing respective roles. One person would be the interviewer with the other taking notes, shifting among these roles in order to ensure objectivity in relation to the informant.

How data were analysed

Data was organised into the FRAM Model Visualizer (FMV) (see Figure 10.2). Notes from each interview were analysed, inspired by an *inductive approach* and *grounded theory*, looking for common patterns in how functions/activities are performed (Kvale, 2002). FRAM Model Visualizer (FMV) was used to visualise and characterise how functions/activities are performed as *work as done* rather than *work as imagined*. FRAM theory advocates a description of the

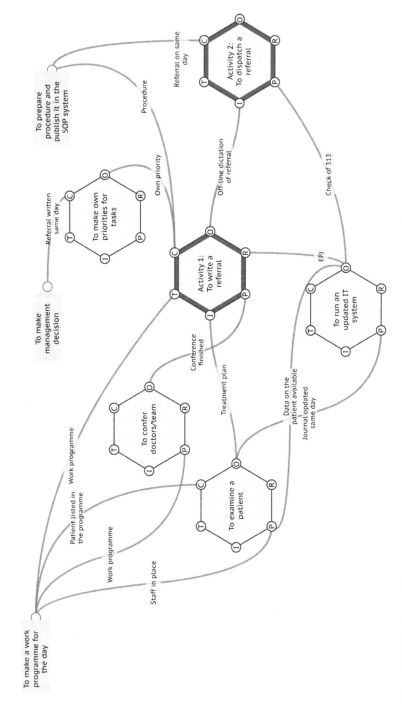

Figure 10.2 FRAM model

patterns for general explanatory statements to describe how work is generally performed. The next step was to look into the different aspects of variability in work processes that may lead to adverse events.

In the FMV, two activities were created: Activity 1: 'To write a referral', Activity 2: 'To dispatch a referral'. New functions were added when needed. The extent for FRAM modelling was set to account for meaningfulness in the interactions, using as much information from the data collection as possible.

Change in work programme

Ahead of activity 1 is an upstream function: 'To examine a patient'. The upstream function for this is: 'To make a work programme for the day'. This function is characterised as a background function because it cannot be changed in its structure but content can change during the day caused by, e.g., acute patients, or illness among staff or patients. Such changes in schedule can cause increased work load and need for conferences. The changes can cause resonance according to the time aspect in activity 1, as 'To make a work programme for the day' becomes a direct upstream function for activity 1. 'The work programme for the day' does not compensate for increased work load, which calls for increased resilience by the staff performing the activity 1.

Incomplete interdisciplinary conference

A precondition for activity 1 is the function 'To confer doctors/team'. With regards to precondition, resonance is also present in the need for conferences as it depends on the presence of relevant professions at the conferences or access in an ad- hoc function. Because of increased work load the clinicians will be occupied elsewhere and cannot participate in the conference. Understaffing and allocation of resources are also a theme.

Disturbance and interruption

Informal conference is then an option for the clinician who needs to confer with another health care professional, but this can cause disruption and disturbance. An informal conference often includes access to the patient journal, MRI scans or other tasks on the computer, which means interruption of ongoing work and switching between patient journals. Activities 1 and 2 are often performed in larger communal offices of six clinicians, and informal conferences will cause disturbance for other colleagues. Disturbance also occurs during supervising the clinical trainees. If the clinicians fail to confer the patient, the task will often be postponed to the next day or week.

Recall systems

Recall systems, e.g., print outs of journals on the desk, writing lists in a diary etc., are relied on when tasks are postponed to the next day or week. Some tasks are clinically triaged as more urgent than others, e.g., a referral for surgery is more urgent than a referral for rehabilitation. This fact will replace the background function: 'To make management decision' with a new function: 'To make own priorities for tasks' (implicit use of recall systems). The new function will occur as a control function for activity 1. The output of the function 'To make management decision' is an instruction regarding all tasks of the day, that they should be finished the same day. All journals and referrals must be dispatched from the Spine Center at the end of the day. The new function can be characterised as a hidden function, leading to a discrepancy in *work-as-imagined* and *work-as-done*.

Activity 2: To dispatch a referral

Activity 1 is an upstream function for activity 2. The activity: 'To dispatch a referral' is not referred to a specific secretary. The activity is performed by all secretaries who are aware of incoming referrals during the day, amidst a lot of other daily tasks. If the secretaries have an increased work load and are inattentive the dispatch can be postponed.

How the analysis results were interpreted and gave rise to recommendations

In the following analysis, the aim was to identify the four fundamental principles of FRAM: (i) principle of equivalence, (ii) principle of adjustment, (iii) principle of emergence and (iv) principle of resonance.

Principle of equivalence

This was not found as no systems break-down or failure were present.

Principle of adjustment

The principle of adjustment is present in the function 'To confer doctors/team'. The clinician adjusts to a new situation when the representation at multidisciplinary conferences is incomplete and hinders completing the task of, e.g., writing a referral. Instead the clinician will seek out an informal conference. This adjustment might be exact, if it is possible to find the appropriate colleagues available later the same day. Generally, an informal conference will be approximate as an adjustment, as it is not always possible to find the relevant clinician and get the necessary information to finish the task the same day. The work programme of the day is not resilient in its structure, and it is not possible to make adjustment in time available for a given task e.g. informal conference. A successful multidisciplinary setting depends on a tight and timely programme, which must be complied with by everyone involved. Delays can occur when there is a complicated patient problem, which in return can result in an incomplete conference. Usually the teams compensate and adjust to each other when a particular profession is missing, because of, e.g., illness or other reasons. There is also an ad hoc doctor on duty every day to adjust for understaffing among doctors.

By contrast, an informal conference is resilient in its need for a conference when needed beyond the ones scheduled. This is regarded as an accepted adjustment and there is a positive culture attached to it, which is stimulated by multidisciplinary interaction in communal offices. On the other hand, this could also cause disturbance and interruption, leading to confusing patient identification, to forgetting tasks and to postponing tasks to another day.

Principle of emergence

If a function cannot be carried out the same day, a successful outcome depends on an adjustment based on the individual mnemonic systems and the individual clinicians' triaged lists of tasks awaiting. The quality of these mnemonic systems is unknown. When a task is postponed there is an increased risk of forgetting information and tasks and confusing different patients, especially in an outpatient setting where the patient is not present as a reminder and source of information. The tasks can be delayed even further in cases when the clinicians are off duty for various reasons. By contrast, the mnemonic systems probably advocate that things go well and the priority list is based on a qualified clinical evaluation. The function 'To make own priority for tasks' is a hidden, new function and has a major impact on *work-as-imagined* represented by the function 'To make management decision'. The new function (implicitly using recall systems) is an adjustment for variability in multiple .functions i.e., regarding the aspect time in the sense of work programme and the aspect precondition. The variability in the new function can be considered one of the reasons for either success or an adverse event.

Principle of resonance

If activity 1 fails it has resonance for activity 2. The consequence of this can potentially be harmful to the patient as further examination or treatment is interrupted.

Results

FRAM has elucidated work conditions and variability, having impact on daily performance. *Time, multidisciplinary conferences* and *disruptions* came up as specific points of interest.

Time

Administrative follow-up after the patients' consultations is postponed to later in the day or the next day. With regards to function 1 this could be postponing writing a referral to, e.g., an MR scan. The staff used their own mnemonic systems and priorities in the administrative follow-ups. This was identified as a hidden function in the FRAM analysis and in contradiction to the specific clinical guidelines. Thus enhancing the risk for oversights and adverse events in patient identification.

Lack of complete multidisciplinary representation at conferences was identified as a factor that could influence the clinical decision making for the patient pathway, and hence writing a referral for MR scan. This finding called for better coordination and back-up in the daily work programmes.

Multidisciplinary conferences

When the time frame was under pressure during the multidisciplinary conferences, it was found that the administrative follow-up was often postponed. Thus, the follow-up was more time consuming than estimated. This enhances the risk of adverse events in administrative patient identification.

Disruptions

Disruptions in the collaborate office environments raised the awareness towards the work culture and common team rules to be managed in the multidisciplinary teams.

Proposing ways to quality improvement

Several reasons for postponing the administrative patient follow-up were identified, causing diverse variability of individual mnemonic systems and priorities. This adjustment is identified as being a catalyst for adverse events with regard to patient identification and oversights. However, whether or not certain mnemonic systems hold a preventive factor to adverse events in a clinical setting where daily adjustments are necessary remains unanswered.

Postponing the patient administrative follow-up is more time consuming compared with completing the full set of tasks on time. This finding addresses the need to focus on how to get all the patient-related administrative follow-up done and how to eliminate counteracting barriers.

After the FRAM project was finished and presented to the staff in the Medical Department, several changes and improvement measures have been initiated.

It was decided to work further with the mnemonic systems as an adjustment factor that could both hold problems and solutions towards patient safety. The main reason being that this finding in the daily routine had never been addressed. Discovering mnemonic systems as a hidden function in the FRAM revealed a large-scale daily variability with a significant systemic impact and this was a new finding. Everybody maintains a personal mnemonic system and much learning in the multidisciplinary team could be drawn from this finding. After a presentation on mnemonic systems, the multidisciplinary teams were given an assignment for a forthcoming team conference. In this they were sharing and discussing advantages and disadvantages of the different mnemonic systems being used in the daily routines. The purpose was to create awareness and assessment of mnemonic systems. The feedback to the Quality and Patient Safety team was very positive.

What actually happened?

The Quality and Patient Safety team found that the findings should be investigated to the next level; e.g., looking closer at mnemonic systems.

But due to strategic re-organisation, the Spine Center of Southern Denmark has undergone several fundamental changes, causing a shift in focus towards other patient safety themes to be addressed. The latest semiannual follow-up on reported adverse events showed new patterns and a significant reduction in adverse events in regards to patient identification and oversights. This cannot be regarded as being a valid reduction of factual adverse advents, as it can be explained as a shift of focus in the clinic due to the ongoing changes.

The findings related to postponing administrative follow-up are a focus of strong interest among both management and clinical staff. This has been discussed and addressed during the organisational changes and is still ongoing. This particular finding is recognised as being very complex and not easy to solve, and so far several measures have been taken in close dialogue between clinical staff and line managers. Some of them being: (i) adjusting daily work programmes to align during the day, (ii) increased specialisation and standardisation in work, (iii) relocation of multidisciplinary team. This particular finding was not surprising but FRAM contributed to the finding being addressed in a patient safety perspective and establishing a collective awareness in the Medical Department.

Conclusions

The FRAM approach has served its purpose for the Spine Center of Southern Denmark:

to elucidate the complexity of a persistent patient safety problem and, in this specific case, describe the functional resonances in a complex work flow that can (i) explain how these adverse events occur in a clinical setting and (ii) suggest new perspectives for improving patient safety.

FRAM has contributed to quality improvement measures that are well-founded in the local context. It supports the view that adverse events should be considered as conditions in working in complex systems to be learned from as opposed to mistakes happening when not following linear processes or guidelines. This key message in FRAM methodology was well received among clinic staff. It was easy to recruit informants for each interview and establish the necessary confidentiality, which is a must for in-depth dialogue about daily practices. Also, the responses to the follow-up initiated by the FRAM project have been quite positive.

To the risk managers it has been essential to participate in the FRAM training course. It has been challenging to change the idea of analysis from a linear mindset to the FMV. The FRAM tool needs an extended introduction to achieve an appropriate understanding of the FRAM method. It has been helpful to be able to combine a training course and an ongoing analysis, enabling questions to be addressed iteratively during the project. It has also been helpful and essential for both a practical and theoretical comprehension of FRAM to be able to discuss the final model and its conclusions with an expert risk manager.

Chapter 11

Engineering resilience in an urban emergency department

Garth Hunte and Julian Marsden

Introduction

In this chapter we describe a case study of *how* an urban emergency department (ED) began a journey to *apply* resilience engineering principles in everyday practice.

Overview

We are undertaking an iterative and emergent practice-based knowledge generation and translation process in which the generic Resilience Analysis Grid (Hollnagel, 2011a) will be adapted, refined and evaluated in an urban ED. We present here our story of how *resilience* offers a lens to consider how we can enhance our collective ability to cope with complexity in everyday practice, and respond in face of the constraints and contradictions of work.

Aim

Our primary aim is to adapt, make operational, and evaluate the Resilience Analysis Grid (Hollnagel, 2011a) for use by emergency care providers, ancillary staff and leaders (ED and organisational) to assess and monitor over time. Our objective is to develop a context-specific framework that has high face and content validity, and is acceptable and feasible for use in a busy urban ED.

A secondary aim is to facilitate a culture of inquiry and wisdom (Reason, 2004) through a series of in-person meetings and online feedback involving questions about how we cope with complexity, and thereby foster colla-borative organisational learning from knowing-*in*-practice (Gherardi and Nicolini, 2003; Orlikowski, 2002) around 'what works'.

Background

Patient safety is a complex (Amalberti et al., 2011) and wicked problem (Braithwaite, Runciman and Merry, 2009). Despite a decade of considerable and diligent effort focused on reducing patient harm, unintended harm remains a pervasive and persistent global health care issue.

Safe practice, however, is not simply a question of eliminating risk. Safety is dynamic (Reason, 1997; Weick, 1987; Weick and Sutcliffe, 2007) and emergent – arising from distributed and collective joint inter-action of care providers, staff and leaders as they cope to sustain performance in the face of risk inherent in clinical work.

Urban EDs are crowded, complex and often brittle high-hazard units where patients are exposed to risk of harm. EDs also epitomise a dynamically inter-active, complex adaptive system that demonstrates considerable resilient capacity. Although EDs have adapted to the problem of over-crowding in a variety of ways, such as using hallways and waiting rooms as treatment spaces, and reactively invoking over-capacity protocols, this adaptive capacity has become strained (Committee on the Future of Emergency Care in the United States Health System, 2007) as overcrowding has increased in severity.

The capability to adapt to meet novel and ambiguous circumstances reflects both the ability to learn and act reflexively (Staber and Sydow, 2002), and requires a degree of 'slack' (Schulman, 1993) to maintain an adequate difference between capacity and task requirement. While an ED may be innately resilient and tolerant of uncertainty, operational performance is interdependent and 'tightly coupled' (Perrow, 1999) with other hospital departments and outside agencies where predictability and regularity may be maximised at the expense of flexibility and expedience. Delays or failures may interact and resonate across the nexus of inter-departmental practices and create unanticipated threats to safety (Waring, McDonald and Harrison, 2006).

Significance

A proactive or anticipatory stance is critical in complex, dynamic high-risk environments in order to respond flexibly and effectively to changing and unknown conditions.

Resilience engineering offers a paradigm for safety management that builds on advances in understanding complex adaptive systems, high-reliability organisations, and how people adapt to cope with complexity in joint cognitive systems to achieve success. This perspective goes beyond reactive risk management of unwanted outcomes to offer a proactive approach to safety that focuses on the ability of a system to adjust its functioning prior to, during, or following events (changes, disturbances and opportunities), and thereby sustain required operations under both expected and unexpected conditions (Wears, Hollnagel and Braithwaite, 2015a). Central to this proactive approach is the understanding that safety is dynamic, emerges from everyday practice, and is something a system *does*.

System resilience links safety to operational performance, and draws attention to how a system anticipates, responds and learns. Emphasis on understanding and increasing what goes right in everyday work leads to improvements in both safety and productivity (Hollnagel, 2009b).

Innovation

The Resilience Analysis Grid has been developed as a methodology for use in risk critical industries (e.g., North Sea Off-Shore Oil production, Australian Radiation Protection and Nuclear Safety Agency, and French Rail) to assess four cornerstone organisational capabilities: the ability to *respond to the actual* – to identify and respond to regular and irregular conditions in a flexible manner; to *monitor the critical* – to monitor short-term developments and threats in order to revise estimates of risk; to *anticipate the potential* – to anticipate long-term threats and opportunities and proactively consider and address them; and to *learn from the factual* – to learn from past events to understand what happened and why.

To date, however, there has been no research on the utility of the Resilience Analysis Grid for safety and performance in health care settings. The Resilience Analysis Grid offers promise as a tool to aid organisations and health care teams to assess their progress in developing a resilient and safe operating environment, and provides a profile of the system that can be used to foster safety as an 'active organizational construct' (Rochlin, 1999).

The generic tool is not intended for use off-the-shelf. Rather, the four sets of questions are intended as a base from which more context specific grids or questions may be developed and tailored (Hollnagel, 2011a). The balance among the four capabilities is relative to the organisation, its goals and situation, and the dynamic characteristics of everyday work.

The purpose of the Resilience Analysis Grid is to provide an effective strategy to rate and monitor the everyday way in which an organisation copes (state of resilience) with the challenges and trade-offs that arise in risk critical work. Repeated application of the Resilience Analysis Grid over time will demonstrate how the profile develops and changes. For example, in the ED context, this could be used to monitor performance measures such as time to physician or mean ED length of stay, as well as strategies to improve capability such as over-capacity and surge protocols.

Methodology

Setting

The study is being conducted at a 440-bed acute care, academic and research hospital located in the downtown core of a major Canadian urban centre. The hospital operates as a publicly funded institution within the Canadian regulatory, economic and socio-political environment, and provides quaternary, tertiary and secondary care to the local community and patients from across the province. It has a longstanding history of providing care to socially and economically disadvantaged populations. The ED has approximately 80,000 visits per year, and is a Level III district trauma centre (no neurosurgical service). Care is provided in five spatially discontinuous treatment areas (six including the waiting room and hall-ways). The department is staffed by nurses, physicians and allied health care workers (unit coordinators, porters, ward aides, technicians, social workers), with support from hospital technicians and clerical staff, and contract services (house-keeping and security).

Unit of analysis

The unit of analysis is the activity system of an historical, socio-technical, and culturally situated hospital ED.

Assumptions and values

The concept of resilience has been explored through the lens of practice theory within the recursive paradigm, recognising multiple versions expressed in participants' perceptions and everyday practices, and consider-ing a collective view as located in situated practice within the historical, socio-technical, and cultural contexts in which interaction occurs, and where practitioners create meaning together.

Participants, recruitment and sampling

Participants came from within the department in leadership, clinical, clerical, technical, support and educational roles, as well as from outside of the department in organizational roles that supported the work of the ED. Relative newcomers and long-standing members were included, as well as those who have had extensive experience in other EDs around the world. Citizen advocate and academic leaders also participated. Sampling was purposive to reflect a broad range of perspectives and voices. The number of participants at each Café ranged between 10 and 18, and over the course of the series of Cafés the total number of participants was 35.

Organisational involvement

Prior to embarking on the process, a steering committee of organisational, academic and patient leaders was created to engage, inform and elicit in kind and financial support. The group met and followed up by email before, during and after data collection.

Dialogue workshop

The content of the domain specific Resilience Analysis Grid is derived from dialogue workshops in a practice community of patients, frontline care providers, support staff and leaders in a hospital ED.

Generic statements from the Resilience Analysis Grid were used as a stimulus to encourage and facilitate participants to discuss system resilience and safety. The tool presents an interpretation of 'resilience', while participants represent their interpretation of it in a semiotic process (Törrönen, 2002).

Face-to-face dialogue workshops were informed by a relational perspective of organisational learning, where learning is conceived as a way of taking part in a social process mediated by artefacts. Workshops were designed in keeping with the non-linear and emergent principles of complexity-oriented models such as the World Café (Anderson, 2011; Brown and Isaacs, 2005; Delaney, Daley and Lajoie, 2006) and Open Space (Owen, 2008) approaches for meaningful engagement within the participatory and relational frameworks of Appreciative Inquiry (Cooperrider and Srivastva, 1987; Grant and Humphries, 2006; Rogers and Fraser, 2003; van der Haar and Hosking, 2004) and action research (Greenfield et al., 2011; Reason and Bradbury, 2001).

The 'Café' strategy enabled conversations to link with and build on others as people moved between groups, fostering dialogue in which the goal was not only thinking together, but also creating actionable knowledge. Seen through the lens of practice theories (Bourdieu, 1977, 1990; Schatzki, Knorr-Cetina and von Savigny, 2001), and situated in a dialogic ontology, 'we' studied 'us' (Kemmis and McTaggart, 2003).

Participants addressed the generic statements for each of the four resilience capabilities, and populated the Resilience Analysis Gridc with examples from the ED context. Participants were then asked to work through the generic questions and re-frame them (greater specificity or quantifiability) as needed to fit their organisational context.

Workshop participants were provided with pens and encouraged to draw, write and record their conversations on the paper tablecloths to capture ideas as they emerged. 'Table hosts' then shared emergent ideas with the large group in an 'open space'.

Each one of five breakfast café workshops opened with a 15-minute presentation and discussion about what resilience engineering is and why it is important, as well as preliminary themes from prior café workshops. An elastic band ball was used as a physical metaphor to help convey the concepts of resilience and brittleness. This was followed by three 20–30-minute café rounds to discuss the generic statements for each capability and

Figure 11.1 Graphical illustration of the final café workshop

their integration, followed by a 30–60-minute open space discussion where emergent ideas were discussed and built upon. Participants were free to record their thoughts and conversation however they liked, including on tabletop paper, sticky notes and poster paper. Graphic illustration of the final café workshop is presented in Figure 11.1.

Data collection

Each dialogue workshop was conducted in a nearby off-site meeting room during the participants' non-working hours. Participants were invited to reflect on and share their overall perceptions of 'resilience' in the ED, and how their work environment helps or hinders them in providing safe care.

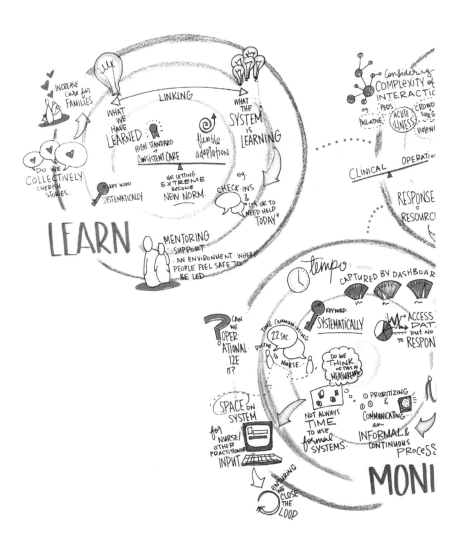

Interpretation

Attention was given to how participants constructed the meaning of action in everyday practice, with reflection on the contrasts and comparisons within and across participants to build interpretations that are grounded in the data and ongoing narrative of professional practice. Inquiry into the experiences and perspectives of emergency health care providers permits description and analysis of the context of everyday operational performance.

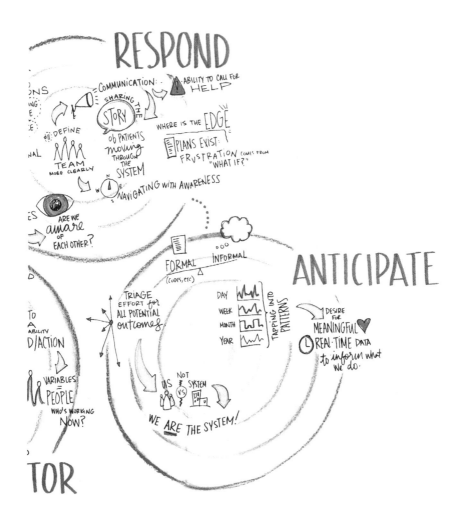

Data analysis

Group transcripts were analysed as a whole unit of discourse. Transcripts and field notes were read and re-read noting narrative chunks, meanings, patterns and themes. Analysis used content analysis techniques and constant comparative methods, with attempts to identify and narrow recurring patterns, look for areas of disjuncture in overall patterns and to refine emergent patterns. Data were coded manually and entered into a computer

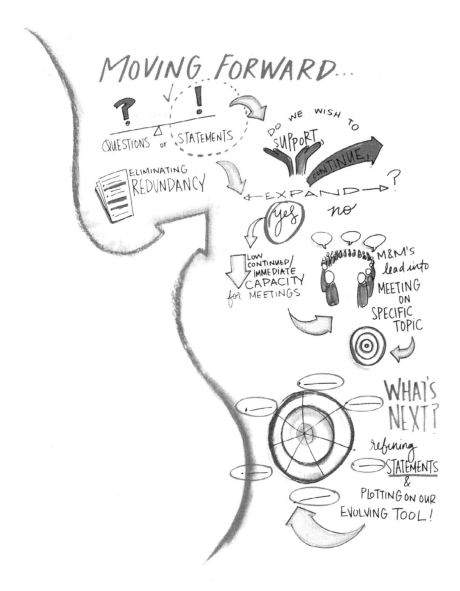

software program (NVivo 10.0) to ease the process of data sorting, storage and retrieval.

Emergent patterns and themes were compared and contrasted looking for discrepancies, coherence and complementarity. The emergency care-specific resilience analysis grid was refined from this synthesis by constructing context-specific sets of statements for each capability.

Findings

Findings presented here include thematic and content analysis from the dialogue café workshops, as well as the first iteration of the context-specific emergency department RAG.

Interaction with the generic Resilience Analysis Grid

The World Café/Open Space strategy was successful in engaging participants to consider the concept of 'resilience' and how it might describe and *apply to* their everyday work. In general, participants engaged more readily with the broad thematic questions for each cornerstone organisational capability than some of the specific questions of the generic RAG. They were confused at times by unfamiliar language and ideas, and struggled to make sense of the generic Resilience Analysis Grid. For example, the statement about an organisational model of the future (ability to anticipate the potential) was met with bemusement. However, through conversation and sharing of stories of everyday practice, they began to populate context-specific examples of how well we as a department and a health care system respond, monitor, anticipate and learn. As they became more at ease with the topic of focus, a playful, yet earnest spirit emerged.

Thematic analysis

Ability to respond

The nature of emergency care is to expect the unexpected, and to be able to cope with 'anyone, anything, anytime' (Zink, 2005). The McGyver/Swiss Army knife concept reflects the inherent adaptability and capability of emergency care providers to cope with and solve complex problems with everyday materials and resources at hand, as in *bricolage*/bric-a-brac work (Lévi-Strauss, 1966) and jazz improvisation. There is a necessary tolerance for risk but the threshold of system decompensation, and the line between success and failure, is often unknown.

Potential events range from the catastrophic and pandemic, to routine everyday events like crowding. Acuity, volume and access are priority events for system response and, as a hub between community and acute care, an ED requires effective interactions across scales, departments and agencies.

These interactions rely on *teaming*, the verb of teamwork (Edmondson, 2012; Nawaz et al., 2014), the *doing* of team in everyday work, which, in part, is dependent on who makes up the *ad hoc* team in the moment, and the resources available (human, equipment, space). Thus, the ability to detect and respond is varied and variable, interpreted and acted upon by team-environment. The ability to work together increases our ability to respond, but this is easier in crisis ('the bomb') than everyday work ('the grind').

Ability to monitor

Overload is a common stress, and has been implicated in the catastrophic collapse of complex systems in multiple domains. The *sine qua non* of an ED is timely response to acute illness or injury. This requires the capability to take on new work. However, as work increases, the mismatch between demand and capacity also increases, and contributes to prolonged length of stay and crowding. This 'falling behind the tempo of operations' is a classic pattern of decompensation, and is a daily pattern in an urban ED. An ED is a complex service environment with many shared resources. System throughput decay caused by sharing is a function of both load on the system and the proportion of service that requires use of shared resources.

An adaptive system is able to cope effectively up to a point (elastic region), gradually starts to fall behind (plastic region), then abruptly slows (fracture or deformation). In a resilient system, large increases in work processed contribute to only small increases in recovery, and the system is able to keep pace. Exploitation of resources (e.g., additional staff, space, support) to enhance slack or capacity for manoeuvre, contributes to improvements in performance by mitigating the disruption, shortening the time to recovery, and increasing the threshold for decompensation.

The goal, therefore, is to have meaningful signals linked to action plans, an adaptive response to degrade gracefully or a transformative response to move to a new state space with different ways of doing things, e.g., the 2011 Stanley Cup Riot (Hunte, 2015). These signals form a suite of operational 'vital signs': metrics that offer an estimate of incoming work, completed work and work in progress, with the purpose of matching capability/capacity to demand in real time. Systematic (re)prioritisation in a dynamic environment helps keep up with current need, but requires flexibility in staffing, 'standards', space and support services to be able to adapt, adjust, distribute, defer or decant work.

Informing data-driven signals are human perceptions of work saturation, the 'Spidey sense' or awareness of danger, which are informal, like expletives and information scraps (e.g., Post-it Notes at the triage desk). Unfortunately, when chronically overwhelmed with complexity or workload, there is often loss of awareness, a decrease in work satisfaction and compassion, leading to conflict and burnout. Maintaining the capacity to match needs and keep up is dependent on organisational support (leadership, financial, operational), as well as orientation to and awareness of what else is happening outside/inside the hospital. As noted previously, this requires effective linkages, communication and attention to cross-scale interactions.

Ability to learn

Organisational learning happens in a shared and contested workplace, where siloing and limited cross over between disciplines reduces opportunities for effective debriefing and collective learning. This suppresses resilience. Moreover, resources for resilience, the unplanned but effective practices that accomplish work safely, are not widely shared. More effort and resources are put into learning from crisis than everyday events. Feedback loops are often not closed, and patients and their families are not routinely invited to contribute what matters to them. Patient outcomes are more than clinical, which calls for human interactions and relationships of empathy and care to avoid vicarious harm from hostility. There can be joy in suffering, meaning making and personal transformation (Frankl, 1985), but is this what the system of health care is learning and bearing witness to?

Ability to anticipate

The rhythmic flow of daily operations, or cadence, is impacted by various drivers and blockers. Using the carousel metaphor (Nugus et al., 2014), the ability to get on, go around, and get off is affected by how many other 'users' are competing for shared finite resources in the moment. Hence, our capability to maintain capacity for manoeuvre in everyday work is constantly negotiated and contested in relationships of power (Hunte and Wears, 2016).

Although the model of the future is unknown, there are repeated patterns of demand based on historical data. The past does not always predict the future, but this information could guide preparation (anticipation) in the long term alongside ongoing reassessment and calibration with real-time data. Anticipation and preparation go beyond the boundaries of an ED, and must involve multi-level stakeholders from community and acute care. Preparation and practice support awareness, skill development and anticipatory actions. For example, regional multi-sector preparations for the 2010 Winter Olympics helped in the 2011 Stanley Cup Riot (Hunte, 2015). Yet, what is also clear from this example is that preparation for the future required a focus, a concerted and organised effort and political will.

Emergency Department Resilience Analysis Grid

There was a preference for statements of agreement over questions, and the ED context-specific Resilience Analysis Grid is presented this way. There are six statements for each capability, edited and iteratively refined from an initial list of 35 statements. Statements were eliminated for clarity and overlap.

Ability to respond to the actual

- We have a list of everyday expected and unexpected clinical, system, environmental and external events for which we prepare and routinely practise action plans.
- We revise our list of events and action plans on a systematic basis.
- We define thresholds to adapt practice and proactively mobilise resources in order to maintain our capacity and capability to response under conditions of increased acuity and volume.
- We effectively communicate and work together as a team within the department, and with other departments and services.
- We have organisational support and resources to maintain our capability to meet acuity and volume demands.
- We link and align our local department adaptations to the organisation and health care system.

Ability to monitor the critical

- We are aware of and attend to formal and informal signals of work saturation.
- We continuously monitor operational tempo (pace) to detect when we are falling behind.
- We systematically and dynamically monitor, (re)prioritise, and match required resources to current patient needs.
- We continuously monitor, measure and update consensus and evidence-based departmental and regional performance metrics and health systems outcomes.
- We regularly monitor system recovery.
- We graphically display critical real-time operational performance indicators (vital signs).

Ability to learn from the factual

- We cherish stories of success and failure in everyday practice.
- We routinely debrief and learn together as providers and supporters of patient care.
- We partner with patients and families to learn what matters to them.
- We support systematic selection, analysis and learning from what happened.

- We follow learning opportunities to confirm that change in action has had the desired effect.
- We share learning across provincial emergency departments, and with the Patient Safety Learning System.

Ability to anticipate the potential

- We make it easy and welcoming to voice potential or anticipated safety threats.
- We routinely anticipate and manage risk of potential or anticipated safety threats.
- We invest in developing and maintaining capability to understand and predict future threats to safety and operational performance.
- We use historical data, early warning signals and thresholds to inform our real-time response.
- We reassess and recalibrate our response based on real-time data. We prepare and practise for potential threats and everyday hazards.

Using the ED-Resilience Analysis Grid

Following development of the initial version of the ED-specific Resilience Analysis Grid through the series of World Café/Open Space dialogue workshops, we initiated a monthly interdisciplinary departmental meeting to consider each domain of the grid, refine statements, survey members of the ED, and address identified gaps. So far, we have mapped the ED resilience profile for the abilities to monitor and respond (see Figures 11.2 and 11.3).

The statements about 'vital signs', thresholds and action plans have been generative. This has led to parallel work on selecting the suite of operational metrics that practitioners and staff look to in order to get a sense of how the department is operating. We have implemented them in a pen and paper version (see Figure 11.4), and an electronic version that is integrated with the electronic bed board is anticipated in the new few months. In addition, consideration of thresholds has led to conversations about flexibility and potential ways of changing our work environment and the way we move patients. We have created a collaborative living document on Google Docs to generate and refine action plans linked to operational thresholds.

Discussion

We have embarked on an organisational learning journey using the generic Resilience Analysis Grid as a stimulus for dialogue and action. The iterative and emergent process of engagement and dialogue with a practice community has been generative. The tool has been edited and refined to create a context-specific framework that has face and content validity, and is

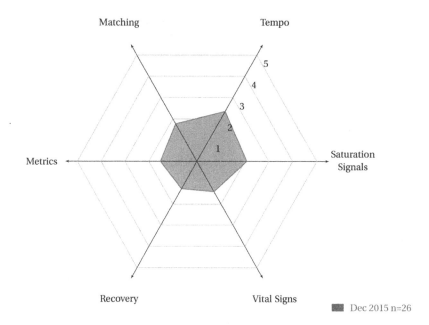

Figure 11.2 The ability to monitor the critical.

acceptable and feasible for use in an urban ED. Even so, the richness and benefit has come as much or more from the process of dialogue with a diverse group of health care system participants, as it has from the product of the adapted grid. The process has been productive in identifying gaps, leading to innovations to facilitate awareness of operational performance, and actions to increase our collective capacity for manoeuvre. We plan to rotate through the ED-specific Resilience Analysis Grid on an annual basis, and iteratively map the resilience profile of the department over time.

Appendix: The Resilience Analysis Grid

Actual

The first set of questions focuses on the *actual*, that is, the ability of the organisation to respond to regular and irregular conditions in an effective, flexible manner.

- **Measuring:** How to respond.
- **Analysis item:** Ability to respond.

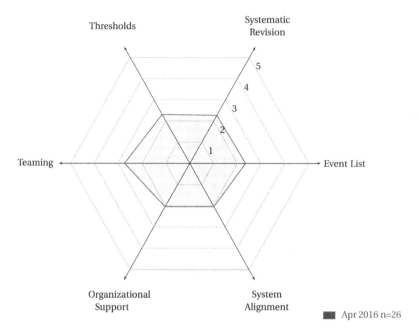

Figure 11.3 The ability to respond to the actual.

- **Event list:** What are the events for which the system has a prepared response?
- **Background:** How were these events selected? (Experience, expertise, risk assessment, or other?)
- **Relevance:** When was the list created? How often is it revised? On which basis is it revised?
- **Threshold:** When is a response activated? What is the triggering criterion or threshold? Is the criterion absolute or does it depend on internal/external factors?
- **Response list:** How was the specific type of response decided? How is it ascertained that it is adequate? (Empirically, or based on analyses or models?)
- **Speed:** How fast is full response capability available?
- **Duration:** For how long can a 100% effective response be sustained?
- **Stop rule:** What is the criterion for returning to a normal state?
- **Response capability:** How many resources are allocated to the response readiness (people, materials)? How many are exclusive for the response potential?
- **Verification:** How is the readiness to respond maintained? How is the readiness to respond verified?

ED Operational Vital Signs

Date/Time (24h clock)

Number of ACUTE patients waiting to be seen
metric of incoming work

Number of ACUTE CTAS 2 patients in an unmonitored space
metric of acute resource matching

Mean ED length of stay
metric of work in progress

Number of patients in RAZ/WR
metric of access block

Total number of ACUTE patients
metric of crowding

Actions

Comments

version 1.1, June 2016

Figure 11.4 Emergency Department operational vital signs (pen and paper version).

Factual

The second set of questions focuses on the *factual*, that is, the ability of the organisation to learn from past events and understand correctly what happened, and why.

- **Measuring:** How to learn.
- **Analysis item:** Ability to learn.
- **Selection criteria:** Which events are investigated and which are not? How is the selection made? Who makes the selection?
- **Learning basis:** Does the organisation try to learn from successes as well as from failures?

- **Classification:** How are events described? How are data collected and categorised?
- **Formalisation:** Are there any formal procedures for investigation and learning?
- **Training:** Is there any formal training or organisational support for investigation and learning?
- **Learning style:** Is learning a continuous or discrete (event driven) activity?
- **Resources:** How many resources are allocated to investigation and learning? Are they adequate?
- **Delay:** What is the delay in reporting and learning? How are the outcomes communicated within and without the organisation?
- **Learning target:** On which level does the learning take effect (individual, collective, organisational)?

Potential

The third set of questions focuses on the *potential*, that is, the ability of an organisation to anticipate long-term threats and opportunities.

- **Measuring:** How to anticipate.
- **Analysis item:** Ability to anticipate.
- **Expertise:** What kind of expertise is relied upon to look into the future (in-house, outsourced)?
- **Frequency:** How often are future threat and opportunities assessed?
- **Communication:** How are the expectations about future events communicated or shared within the organisation?
- **Strategy:** Does the organisation have a clearly formulated model of the future?
- **Model:** Is the model explicit or implicit? Qualitative or quantitative?
- **Time horizon:** How far ahead does the organisation plan? Is the time horizon different for business and safety?
- **Acceptability:** Which risks are considered acceptable and which unacceptable? On which basis?
- **Aetiology:** What is the assumed nature of future threats? Same as previous threats/accidents? Combination/extrapolation of known accidents/incidents? Completely novel threats?
- **Culture:** Is risk awareness part of the organisational culture?

Critical

The fourth set of questions focus on the *critical*, that is, the ability of the organisation to monitor short-term developments and threats and revise risk models.

- **Measuring:** How to monitor.
- **Analysis item:** Ability to monitor.
- **Indicator list:** How have the indicators been defined (by analysis, by tradition, by industry consensus, by the regulator, by international standards, or other)?
- **Relevance:** How often is the list of indicators revised, and on what basis?
- **Indicator type:** How many of the indicators are leading, and how many are lagging?
- **Validity:** For leading indicators, how is their validity established?
- **Delay:** For lagging indicators, how long is the lag?
- **Measurement type:** What is the nature of the 'measurements'? Qualitative or quantitative? (If quantitative, what kind of scaling is used?)
- **Measurement frequency:** How often are the measurements made (continuously, regularly, every now and then)?
- **Analysis:** What is the delay between measurement and analysis/interpretation? How many of the measurements are directly meaningful and how many require analysis of some kind?
- **Stability:** Are the effects measured transient or permanent?
- **Organisational support:** Is there a regular scheme or schedule? Is it properly resourced?

Chapter 12

Patterns of adaptive behaviour and adjustments in performance in response to authoritative safety pressure regarding the handling of KCl concentrate solutions

Kazue Nakajima and Harumi Kitamura

Introduction

Since health care is a complex and dynamic work environment, everyday clinical work must be flexible rather than strictly controlled. In fact, recent pressures to improve patient safety have led various authoritative institutions (e.g., governments, accreditation bodies and academic societies) to issue safety alerts or recommendations to eliminate what goes wrong rather than to facilitate what goes right. To promote resilient health care, it is essential to understand how health care professionals actually work in a given environment. One way to understand everyday clinical work is based on the concepts of work-as-imagined and work-as-done. Study of everyday clinical work is to find out what happens and how something happens instead of why it happened; in other words, to identify patterns in individual and collective behaviour and the determinants of behaviour patterns. In this chapter, using the example of handling potassium chloride (KCl) concentrate solutions as a significant issue in patient safety, we report how Japanese clinicians have handled KCl concentrate solutions (i.e., work-as-done) under the safety policy recommended by authoritative institutions (work-as-imagined), examining at patterns of collective and individual behaviour of clinicians and their determinants.

Rationale for the data collection approach

Overview of problems and safety measures related to KCl concentrate solutions

Potassium chloride concentrate solutions are used to correct low potassium levels, which can be life-threatening and require urgent treatment. These solutions are high-risk medications that can be fatal if given inappropriately. Fatal consequences associated with inappropriate rapid injection of

KCl concentrate solutions stocked in patient wards, usually in ampoule form, have been reported in multiple countries. Patient safety alerts that recommend protocols for safer handling of KCl concentrate solutions in hospitals have been issued (Joint Commission on Accreditation of Healthcare Organizations, 1998; National Patient Safety Agency, 2002; Australian Council for Safety and Quality in Healthcare Medication Safety Taskforce, 2003). These alerts emphasise two points: removal of ampoules of KCl concentrate solutions from ward storage in non-critical care areas and substitution of commercially premixed solutions, in order to reduce the risk of accidental rapid injection; and restrictions on the storage of ampoules in critical care units, along with strict safety procedures and staff education, to ensure that seriously ill patients promptly receive high concentrations and doses of KCl.

Necessity for information about work-as-done with regards to handling KCl concentrate solutions

The Patient Safety Promotion Committee of the Japan Council for Quality Health Care (JCQHC), a hospital accreditation body, released their first alert in December 2003. It recommended removal of ampoule-type KCl concentrate solutions from patient care areas, except for specific units authorised by hospital policies. This strategy is similar to approaches adopted by other countries. Six months later, however, the Committee issued a second alert with an aggressive recommendation for total removal of ampoule-type KCl products from all areas of the hospital other than pharmacy department, including critical care units. The rapid shift towards a stricter policy in a short period of time can be attributed to the continued occurrence of fatal events with the initially recommended policy. The second alert also recommended the use of KCl concentrate solutions in prefilled syringes (20 mmol potassium/20 mL), which became commercially available after repeated incidents. These syringes are designed to be connectable only to fluid infusion bags or bottles to achieve dilute potassium concentrations of less than 40 mmol/L; they are not directly connectable to venous lines. This policy raised clinical concerns from physicians, particularly in cardiology, cardiovascular surgery and critical care, about the potential inability to promptly administer concentrated KCl solutions to patients when needed. Several follow-up alerts from the regulatory body (Ministry of Health, Labour and Welfare, 2004) and academic societies with slightly different interpretations of the JCQHC alert (Japanese Board of Cardiovascular Surgery, 2004; Japanese Circulation Society and six other cardiology-related societies, 2004) sought to allay these concerns.

In 2015, JCQHC's Adverse Event Prevention Division, the centre that runs the national adverse event reporting system, issued a patient safety report about a case that involved a unique way of preparing medication using a prefilled syringe-type KCl product. A nurse drew a KCl concentrate solution from the prefilled syringe product into another empty syringe by sticking the needle of the empty syringe into the prefilled one. Subsequently, a resident physician then injected the undiluted solution rapidly through a peripheral venous line.

This event suggests that there are differences between the JCQHC's second recommendation (work-as-imagined) and actual practice (work-as-done) in handling KCl concentrate solutions. Such differences might have be fulfilled by clinicians' adjustments, as evidenced by the surprising invention of a risky handling technique for safer products of KCl concentrate solutions. To understand work-as-done for handling KCl concentrate solutions in Japanese hospitals, we needed the following information: (1) rates of nationwide hospital compliance with the JCQHC's safety recommendations; (2) reasons for barriers to complying with the JCQHC's second recommendation; (3) options and adjustment in handling of KCl concentrate solutions in response to the JCQHC's second recommendation.

How data sources were selected

We selected data sources that could directly and indirectly reflect 'work-as-done' for handling KCl concentrate solutions as described above. Table 12.1 shows the data sources chosen for the study. First, data on nationwide compliance data with JCQHC recommendations were obtained from JCQHC surveys of accredited hospitals regarding the handling of medications containing concentrated potassium, which were published in JCQHC journals. The first survey was sent to 755 accredited hospitals in 2004, 6 months after the first alert. The second one was sent to 1,258 hospitals from late 2006 to early 2007, approximate 2 years after the second alert. The number of participating hospitals in both surveys seemed sufficient to represent typical responses of hospitals to the recommendations.

Second, to understand reasons for barriers to comply with the second recommendation of total removal of ampoule-type KCl from all patient care areas, we chose data sources from our hospital (Osaka University Hospital with 1,074 beds) because our hospital had approved three critical care areas (including the Intensive Care Unit (ICU)) to stock and use ampoule-type KCl concentrate solutions for the past 10 years for clinical reasons. Data sources included documents from the hospital's committees, including discussions relevant to the handling of KCl concentrate solutions and interviews with clinicians who were knowledgeable about treatment of hypokalemia and handling of KCl products.

Third, to capture adaption and adjustments made by hospitals in response to the second recommendation, medication supply data, observation of actual

Table 12.1 Data sources used to understand 'work-as-done' with regards to the handling of KCl concentrate solutions

	Data sources	Information contained
Nationwide data	JCQHC surveys results for accredited hospitals	Nationwide compliance with JCQHC's first and second recommendations
Data from Osaka University Hospital	Minutes and memorandums of hospital committees	Reasons for needing to stock ampoule-type KCl concentrate solutions in critical care areas
	Medication supply data for KCl concentrate solutions	Compliance with the current hospital policy
	Observations of medication practices by ward pharmacists	Compliance with the current hospital policyAdjustment made when handling of KCl concentrate solutions
	Interviews with clinicians	Reasons why stocking ampoule-type KCl concentrate solutions is necessary in critical care areas Options and adjustment in response to JCQHC's second recommendation
Data from other hospitals	Mailing list of Japanese critical care physicians	Reasons for difficulties when ampoule-type KCl concentrate solutions were not stocked in critical care units Options and adjustment in critical care units in response to JCQHC's second recommendation
	Interviews with patient safety officers and managers	Options and adjustment in critical care units in response to JCQHC's second recommendation

medication practices by ward pharmacists and interviews with clinicians in our hospital were used. Expert opinions of Japanese critical care physicians and interviews with patient safety officers and managers of other hospitals with similar characteristics of ours were selected as external data sources.

How data were collected

Data collection was conducted as part of patient safety activities in our hospital in 2015 as a review of established policy on KCl concentrate solutions in response to JCQHC's second recommendation. Therefore, clinical staff working on patient safety at our hospital could easily gain access to in-hospital data sources. Information from other hospitals was readily gathered from the network of Japanese national university hospitals for patient safety.

Minutes and memorandum from our hospital committees: Clinical staff at our hospital obtained information about the necessity of ampoule-type KCl concentrate solutions and stocking of these items in patient care areas from the minutes and memorandums of our hospital committees, including the Patient Safety Committee, Pharmaceutical Committee, and Clinical Service Committee. These included results of a survey of all inpatient wards and units of the hospital aimed at identifying measures to be taken in response to JCQHC's first recommendation.

Interviews with clinicians at our hospital: Our hospital's Patient safety officers interviewed senior physicians and nurse managers responsible for patient safety in their wards to determine what adjustments were needed and what alternatives have been proposed for the hospital to follow the second recommendation.

Medication supply data: Medication supply data were extracted from hospital information systems to check whether ampoule-type KCl concentrate solutions had been delivered to the places that the hospital policy approved.

Observation of medication practice at inpatient areas: Information about actual handling of KCl concentrate solutions was obtained from ward pharmacists who had the opportunity to directly observe medication practices of physicians and nurses in patient care areas.

Mailing list of critical care physicians: Data about adaptation and adjustments were gathered from an open mailing list of Japanese critical care physicians where they discussed how to promptly correct low potassium levels in ICU patients in the new safety environment after JCQHC's second recommendation was issued.

Interviews with patient safety officers and managers of other hospitals: We asked patient safety officers and managers of four other similar university hospitals about their policies and actual practice related to medications containing concentrated potassium as of 2015 by phone or email.

How data were analysed

Nationwide compliance with the authoritative recommendations

According to JCQHC's first survey in 2004, more than 300 accredited hospitals stored ampoules of KCl concentrate solutions. Sixty per cent stored ampoule-type KCl products in patient care areas. Although the rest of the hospitals stocked the ampoule-type products only in pharmacy departments, approximately 60 per cent of these institutions released these products to patient care areas without dilution, resulting in the temporary presence of KCl concentrate solutions in patient care areas. The second survey revealed that, as of early 2007, more than 400 accredited hospitals needed ampoules of KCl concentrate solutions in any particular location within the hospital. Despite JCQHC's second recommendation that ampoule-type KCl products be removed from all areas other than pharmacy departments, approximately 170 hospitals still stocked them in patient care areas, citing clinical necessity. These areas included ICUs, operating rooms (ORs), emergency rooms (ERs), coronary care units (CCUs), high care units (HCUs), neonatal intensive care units (NICUs) and general wards. Prefilled syringe-type KCl concentrate solutions were implemented in almost half of the responding hospitals.

JCQHC's nationwide survey showed that substantial numbers of hospitals needed to store ampoules of KCl concentrate solutions in specific patient care areas, contradicting the external safety recommendations. At the same time, many hospitals shifted from conventional ampoule-type products to prefilled syringe-type products designed to avoid rapid injection of concentrated solutions.

Reasons for barriers to following the recommendations

The committees' minutes and memorandums and interviews with physicians in our hospital uncovered two major reasons why it was necessary for ampoules of KCl concentrate solutions to be stocked in wards. First, concentrated solutions of KCl ampoules were needed because patients in critical condition with hypokalemia, e.g., patients with severe heart failure, often require strict fluid volume control. Second, hypokalemia requires prompt correction because it may cause life-threatening arrhythmia. If ampoule-type products were stored in the pharmacy department, immediate delivery of medications to a unit would be difficult or impossible. For example, on holidays or after business hours clinicians had to go to the pharmacy department to pick up urgently needed medications because of the lack of a just-in-time medication delivery system. One nurse related an experience in which it took 40 minutes for her to obtain a prescribed medication during a night shift.

Our closed ICU, which is staffed by intensivists, uses large volumes of KCl concentrate solutions (700–1200 ampoules per month). More than 60 per cent of patients are cardiovascular surgical patients and they have ever-changing conditions. If the hospital followed the recommendation, a large number of ampoules would be delivered from the pharmacy department according to physicians' orders. Alternatively, KCl in prefilled syringes would be used in dilutions of less than 40 mmol/L, as prescribed in the drug package insert. The results showed that compliance with the external safety recommendations would endanger patients' lives and threaten the efficiency of clinicians in their specific work environments.

Options and adjustments in the critical care environment

Critical care physicians on the mailing list discussed how they provided high concentrations and doses of KCl solutions through central venous lines using syringe pumps, after JCQHC's second recommendation. One option was to stock ampoule-type products in their ICUs. Another option was to draw concentrated solutions from prefilled syringes of KCl into other syringes by sticking the needle of an empty syringe into the prefilled syringe.

Medication supply data from the hospital information system confirmed that ampoule-type products had been released only to authorised areas. On the other hand, ward pharmacists reported observing adjustments (drawing concentrated KCl solutions from prefilled syringe) in the CCU and cardiovascular care unit (CVCU). Both units were recently opened in one corner of the regular wards, where conventional hospital policy specified that only prefilled syringe-type products could be used. Senior physicians in both units claimed that ampoule-type products must be available because of medical necessity. One nurse manager wanted to use ampoules of KCl solutions instead of prefilled syringes because they experienced technical difficulties in drawing solutions from prefilled syringes. Interviews with patient safety officers and managers from four other university hospitals also revealed that all of the hospitals adopted only prefilled syringe-type KCl products, and permitting the drawing of concentrated solutions from prefilled syringes in specific critical care areas varied among hospitals.

Summary of the gap between work-as-imagined and work-as-done

Table 12.2 summarises the gap between work-as-imagined of the authoritative institutions and work-as-done in critical care settings, with regard to the handling of KCl concentrate solutions. From the outset, however, it was clear that the objectives of the authoritative institutions and clinicians differed. External safety organisations sought to eliminate rare but fatal adverse events due to rapid injections of ampoule-type KCl products in any patient care area. Meanwhile, the clinicians' work was aimed at saving lives by administering concentrated solutions of KCl. The goals of actions taken were also divergent. The safety institutions tried to remove ampoules, a source of fatal medications, from all patient care areas and replace them with safer and more dilute products, whereas clinicians attempted to obtain concentrated solutions of KCl for seriously ill patients by any efficient means available within the system (e.g., the hospital) or the subsystem (e.g.,the ICU) where they work. Hospitals commonly adopted one of two options: stocking of ampoule products in critical care areas, against the safety recommendations; or implementation of prefilled syringes, consistent with the recommendation.

Table 12.2 Gap between work-as-imagined and work-as-done with regard to the handling of KCl concentrate solutions

	Work-as-imagined(by the authoritative institutions)	Work-as-done(in settings with seriously ill patients)
Purpose of work	To eliminate rare but fatal adverse events due to rapid injections of KCl concentrate solutions	To save patients' lives by providing concentrated solutions of KCl to specific patients who require prompt correction of hypokalemia.
Goal of action taken	To remove ampoules of KCl concentrate solutions from patient care areas and use diluted solutions for patient safety	To obtain concentrated solutions promptly and efficiently for patient treatment
Means	1. Store ampoule-type concentration solutions in pharmacy departments and deliver them to patient wards when needed	1. Continue storing ampoule type KCl products
	2. Use prefilled syringe products with error-proof design in dilutions with potassium concentrations of less than 40 mmol/L	2. Adopt prefilled syringe-type KCl products and draw concentrated solutions form prefilled syringes into an empty syringe

In critical care settings with only prefilled syringe products of KCl, clinicians made the adjustment of drawing concentrated solutions of KCl from the prefilled syringes. Use of prefilled syringe-type KCl did not always mean that diluted solutions were prepared for infusion in the prescribed way.

How the analysis results were interpreted and gave rise to recommendations

Patterns emerging from simple rules guiding clinicians' behaviour

In complex adaptive systems, it is known that certain patterns can emerge from interactions of agents in the system based on simple behavioural rules (Paries, 2006). This study illuminated two typical patterns in the collective behaviour of clinicians and hospitals in response to external safety pressure: to continue stocking ampoules of KCl concentrate solutions in their units against the recommendation (denial); or to stock and use prefilled syringes of KCl according to the recommendation (compromise). These patterns seemed to emerge from three simple rules guiding clinicians' behaviour in interacting with external perturbations 1) to maximise patient benefit; 2) to pursue work efficiency; 3) to resolve problems within their authority.

In the compromise pattern, adjustments are necessary to fill in the gap between current and new practice (Hollnagel, 2015c). An adjustment is a solution that works under the conditions of bounded rationality, i.e., satisficing (Simon, 1956). This usually involves substituting means to reach a subsidiary goal (Hollnagel, 2015c). In treating critically ill patients with hypokalemia, the subsidiary goal is to obtain concentrated solutions of KCl. Therefore, in settings with only prefilled syringes of KCl, drawing concentrated solutions of KCl from prefilled syringes is an indispensable adjustment, although these products are supposed to be used at considerably greater dilutions.

Potential risks of adjustments at the sharp end

Adjustments make everyday clinical work successful, but at the same time pose a risk of adverse outcomes. Once a concentrated solution of KCl is drawn from a prefilled syringe product into another syringe, the wrong medication could be administered due to confusion with other medications in similar syringes. Since syringes can be connected to three-way stopcocks, rapid injection of the concentrated solution could occur. When such an adjustment becomes habituated, syringes designed to error-proof are no longer safer, because they are treated as containers of KCl concentrate solutions, just like ampoules. The case of rapid injection using a prefilled syringe-type KCl product described above illustrates that this concern has become a reality. In the most recent alert, JCQHC's Adverse Event Prevention Division banned the drawing of KCl concentrate solutions from prefilled syringes. This recommendation may require clinicians to invent another surprising adjustment to obtain KCl concentrate solutions for patients with hypokalemia. Moreover, in a complex adaptive system, interactions among various adjustments across countless clinical procedures other than handling KCl solutions may also play a role in the occurrence of unexpected outcomes though functional resonance (Hollnagel, 2012). Given the potential risks of adjustments at the sharp end, adjustments cannot simply be left as they are.

Necessity for a systemic approach

In order to facilitate the safety and efficiency of everyday clinical work, we need to reduce unnecessary adjustments at the sharp end. A systemic approach should be taken; otherwise health care professionals or institutions will tend to seek quick, realistic solutions within local systems to which they belong and in which they have authority. First, a one-size-fits-all approach is not preferred, because clinical context differs by setting. Instead, hospitals should be given discretion regarding resource allocations (e.g., centralisation of patients in critical condition in the ICU, intensified staff education regarding medications, involvement of pharmacists in medication preparation in patient wards and improvements in the medication delivery system) so that they can perform resiliently. Second, we should be cautious about frequent or additional changes in policies every time a new problem occurs, because we do not have sufficient knowledge about their consequences (Hollnagel, 2015c). Finally, a systemic approach should include innovations such as development of ready-to-use concentrated solutions of KCl for administration via central venous lines, hopefully designed in a manner that prevents rapid injection and cognitive mix-ups.

Conclusion

We illustrated the adaptive behaviour of Japanese clinicians and hospitals in response to authoritative safety recommendations regarding the handling of KCl concentrate solutions. Patterns of adaptive behaviour (denial and compromise) were based on simple rules: maximising patient benefit, pursuing work efficiency and fixing problems in local systems within one's authority. Based on these findings, we conclude that when a change is implemented in the system it is necessary to carefully monitor what happens at the sharp end, learn how it happens, anticipate what will happen in the future and be ready to respond from a system-wide perspective.

Chapter 13

A case study of resilience in inpatient diabetes care

Alistair Ross, Janet E. Anderson, Alison Cox and Rifat Malik

Introduction

Diabetes inpatient care

Inpatient care for people with diabetes in the United Kingdom's National Health Service (NHS) is important, with around 15 per cent of all patients having the condition, increasing the risk of long hospital stays and readmissions when compared with the general inpatient population (Smith et al., 2009). There is also a cost burden of some £1.5m per hour/or 10 per cent of total NHS expenditure (Diabetes UK, 2010).

Simplistically, good care is defined as the avoidance of two conditions at opposite ends of a spectrum of glycaemic control: Life threatening levels of high or low blood glucose (Diabetic Ketoacidosis – DKA), associated in Type 1 diabetes with lack of insulin and high levels of ketones and in Type 2 diabetes with Hyperosmolar Hyperglycaemic State (HHS), and Hypoglycaemia, a low blood glucose, which the patient is unable to treat themselves therefore becoming at risk of coma, associated with both Type 1 and 2 diabetes. Initial assessment and treatment planning, followed by education, medication and nutrition management (in partnership with patient self-management skills) are key to maintaining optimal levels of blood glucose concentration (around 3.5–9 mmol/l) and avoiding these complications.

Improvement efforts

Concerns about the standard of inpatient diabetes care are widespread (Diabetes UK, 2009) and despite a plethora of statements and audits improvement has been slow (Clement et al., 2004). Standardisation of treatment in accordance with clinical guidance and procedural instruction for staff are, as might be expected, one of the key quality and improvement strategies (Roman and Chassin, 2001; Reynolds, 2007). However, there are many interacting factors that make this less than straightforward. Unthinking compliance with protocol may thus be in itself a risk factor, in a complex environment of multidisciplinary working, handovers and communication, variable insulin technology, and multiple patient factors including adherence and self-management, co-morbid conditions and polypharmacy (Grant et al., 2004).

Specialist care teams

Partly in response to this complexity, a significant proportion of NHS acute trusts in the UK now operate a dedicated specialist team for diabetes inpatients. Most diabetic inpatients are admitted for a condition not related to their diabetes control and therefore a myriad of teams throughout the hospital (for example those in charge of patients in surgical or elderly care wards) need to liaise with diabetes specialists in proving care. As well as the safe glucose levels mentioned above, good diabetes care must therefore also not interfere in the resolution of the primary admitting complaint.

Specialist teams usually comprise of a mix of specialist physicians, nurses, podiatrists and dietitians, and can include others such as clinical psychologists. There is some evidence that length of stay and associated costs can be reduced by access to such a service (Kerr, 2011). Further work shows an increase in patient awareness (Davies et al., 2001) and a reduction in medication prescribing errors (Courtenay et al., 2007).

Resilience Engineering

One of the known problems in health care research, as opposed to medical research, is that the linear assumptions inherent in evidence-based medicine (studying closed systems in the physical world) vanish to a great extent when exploring how communities of people interact and behave in social situations to deliver care (Marchal et al., 2013). Here, the number of mechanisms and potential paths to outcomes means that controlling for all relevant interacting and confounding variables is not possible. Contextual factors are also important in shaping how people work and so the very factors that might be controlled in other domains become important explanatory mechanisms themselves.

A basic premise of Resilience Engineering (RE) is that outcomes arise from a fluid arrangement of system components (Hollnagel, Woods and Leveson, 2006). Resilience Engineering (RE) in health care brings together a diverse but interrelated series of concerns with prevailing approaches to health care improvement, including: the over-extension of biomedical models preoccupied with linear effects; an over-reliance on reactive incident analysis, the ubiquity of 'human error' as an explanation for unwanted outcomes; and the lack of a theoretical model about how all the parts of systems interact. In socio-technical systems people both act upon, and adjust behaviour in response to, these dynamic structures, individually and collectively. A prerequisite for improving outcomes is a good understanding of the ways in which system elements are related, which can be used to understand and predict the effects of changes in processes. Importantly, systems self-regulate; that is certain states will be arrived at, following from initial conditions, through dynamic feedback loops that may or may not be formal, or visible to observers. An important task is thus to study responding, monitoring, anticipation and learning (Hollnagel, 2011a) at all levels.

The CARE model for applying Resilience Engineering

Anderson et al. (2014) outline a theoretical model that was developed as a guide to applying RE in health care. The Concepts for Applying Resilience Engineering (CARE) model provides a framework for studying how organisational resilience is manifest in health care, how it contributes to outcomes and how it might be strengthened. The model, shown as Figure 4.1 in Chapter 4, proposes that 'work as imagined' (Debono and Braithwaite, 2015) can be conceptualised as an ideal or designed alignment between demand and capacity. Work is designed around the notion that capacity is provided to meet projected demand. Procedures and organisational functions are assumed to be capable of maintaining this relationship, for example by limiting services when demand falls (e.g., skeleton staff at weekends), calling for expert help (when complications arise), procuring new equipment (when technological advances arise) and training people to ensure suitable qualified and experienced staff.

The key to the model is the inbuilt contention that demand and capacity are *always misaligned in the real world*. The degree may vary, but the clinical demands and the work environment are too complex, and there are too many possible demand-capacity interactions, for the two to ever match completely. Adjustments, by which clinicians work around problems and devise solutions, are central, and outcomes are thus linked to a) variable demands, b) capacity to meet these (which also varies), and formal and informal attempts at different organisational levels to adjust to these multiple dynamic relationships.

Aims and objectives

The main aim of the case study described in this chapter was to use the CARE model to explore, from a Resilience Engineering perspective, what happens in the clinical micro-system of inpatient diabetes care, and how Ross et al. (2014) previously reported a concept map (Novak and Canas, 2006) of the diabetes work system, which represents the factors involved in staff (diabetic specialists and acute medicine non-specialists) decision making. In this chapter we further analyse these interview data using the CARE conceptual model.

Specific objectives were:

- To describe the delivery of care as it works in practice.
- To illuminate misaligned demands and capacities that affect day-to-day work.
- To investigate the adjustments carried out in response to variable conditions and how these relate to outcomes.

Case study methods

Rationale for the approach taken

The model provided a theoretical framework to guide this study of resilience in the inpatient diabetes care process. The model contains various implicit hypotheses about resilience mechanisms and was used to direct attention to misalignments between demand and capacity, adjustments to these, and how these link to outcomes, both acceptable and unacceptable.

Applying such a model requires that researchers gain an understanding of everyday clinical work, which in turn presupposes qualitative or case study methods (Robson, 2015) to gain familiarity with the environment. We attempted to 'enter the social reality of diabetes care' through close engagement with the care teams, including being briefed by specialists, having regular meetings to discuss audit and incident data, attending team meetings, and then observing and further engaging with staff on the wards.

Data sources

The study took place on two adult general acute admission wards, which provide medical care for up to 72 hours after admission. The diabetes specialist team provides proactive review of all diabetic patients on these wards and, therefore, interaction between ward staff and the specialist team occurred frequently. Staff were selected for interview purposively. They did not have to be specialists, but they had to at least interact with diabetes patients and/or specialist teams to be eligible. In the event, no one refused interview, and we stopped at 32 in-depth interviews once a) data saturation had been achieved and b) a range of staff at all levels had been consulted.

Data collection

The interview process was based on the assumption that those working in a system are the best source of information about how work is accomplished, and that this information can be elicited using cognitive task analysis (CTA) techniques (Hoffman and Militello, 2008). CTA refers to a suite of related methods that are designed to capture the nature of the work performed by experts in complex and cognitively demanding domains. The Critical Decision Method (Hoffman, Crandall and Shadbolt, 1998) is one such method. It is a knowledge elicitation technique, usually applied retrospectively, which is designed to gather information about critical events and decisions.

It has been a tenet of safety research for decades (Flanagan, 1954) that the non-routine events (critical, and often negative) are the 'richest source of data about the capabilities of highly skilled personnel' (Klein, Calderwood and MacGregor, 1989: 465).There are many reasons for this bias. Underlying cognitive models imply or assume that problem solving (knowledge-based behaviour) is mainly of interest as it applies to unusual or unexpected conditions. Routine events are assumed to be largely covered by algorithm (rule-based behaviour) or by automatic processes that warrant little explanation. In addition, general attributional biases towards explaining the negative, and regulatory requirements to do so, combine to give a powerful pull towards critical event analysis (Wallace and Ross, 2006). But uncertainty, the bedrock of decision, is also present in *everyday and routine events and working practices* in complex work domains, due to the misalignments and variances abstracted in the CARE model.

We thus adapted an interview schedule drawn from general decision elicitation techniques so that the routine adjustments necessary for care could be studied. The Critical Decision Method (CDM) was adapted to focus first on how *work is normally accomplished*. Non-routine or exceptional events were then discussed in the same way. Prompts were: What were the main goals and objectives in relation to inpatient diabetes? What challenges arose? How were these responded to and/or overcome? What courses of actions lead to what outcomes?

Data analysis and interpretation

Interview transcripts were imported to QSR NVivo 10.0 and analysed using iterative qualitative analysis techniques described elsewhere (which have been used for major safety work before (Wallace, Ross and Davies, 2003)). The elements of the model were used to identify the high-level concepts of demands, capacities, adjustments and outcomes. We then analysed the data to identify resilience trajectories involving couplings and interactions.

Results

Organisation of care

The specialist inpatient diabetes team was originally introduced to provide expert input for difficult cases. The role also encompassed a) patient education and b) education on diabetes care for the multidisciplinary ward team. After an initial consultation by the team, everyday care is carried out by nurses on the ward. This is reflected in accounts such as this from one of the non-specialist doctors:

> So routine diabetics, we would just get the diabetes nurse to see about education and follow up and get them home. I think the diabetes, especially by itself, say if it's a newly diagnosed case, they will want the registrar or consultant to see, or of course they're more complicated. But otherwise, most will get seen by the diabetes nurse. They might need minor changes in their doses or medication and then they will go home. So, that's how we handle that. (Doctor, ward)

Note how various adjustments are part of designed capacity: whether to see a specialist consultant, and whether to adjust dose/medication, are aspects of care for which various contingencies exist. Challenging cases typically involve the team being called to adjust or review the treatment plan – this is typically due to the clinical complexity of the case or social issues such as patient non-compliance (rejection of medical advice), language difficulties etc. Where such clinical complexity is encountered, specialist involvement is more likely:

> [P]atients that [. . .] have complex conditions which cause their diabetes to be quite challenging. They're generally the patients that I see. (Diabetes Specialist Nurse)

Essentially, the specialist team is intended as a problem-solving function, to be initiated in cases where the care moves beyond simple insulin/nutrition regulation that can maintain blood sugar levels within optimal levels:

> So it's a bit like a jigsaw puzzle and how you put it, all the pieces together. (Diabetes Specialist Nurse)

> [I]t's absolutely multidisciplinary [. . .] it involves the diabetic foot practitioner when they go home [. . .] and all this has to be arranged. The GP has to know about what has happened. (Junior doctor)

This model of delivery (ward nurses follow protocol, specialists solve complexity) is fairly standard in all health care specialties and most people describe how this can work to patient benefit. There are many instances reported of just such a delegation of labour, leading to goals being achieved:

> I think for complicated problems, so like the second case, then the knowledge we bring to this situation is not something that will be shared amongst other teams, other special teams, they don't know about

the different types of insulin and they won't necessarily pick out why sugar control is erratic, or they won't necessarily be able to come up with a target for blood sugars [. . .] So that's where we're quite useful. I think we're quite useful in terms of organising the care in general, both in hospital and on discharge. And for the routine cases [. . .] double-check that they're prescribing things at the right times of day or prescribing the right type of insulin at the right time of day. And then if they've got erratic blood sugar control, then that's when we become useful again. (Diabetes Specialist Nurse)

The decision-support function of a team with a wealth of experience of complex cases, helping non-specialists and giving advice to patients, is clearly felt to reduce complications, minimise risk of infection, facilitate a shorter length of inpatient stay and minimise risk of readmission. In summary, *demand* (daily arrival of inpatient diabetes patients, some with complex comorbid conditions) is intended to be met by *capacity* consisting of three main elements:

1   A ward team with the skills to manage basic insulin regulation and nutritional care.
2   A specialist team to:

   a   Assess on arrival.
   b   Educate patients (including newly diagnosed).
   c   See complex cases or those requiring a review of treatment.

3   A structure for multidisciplinary interaction, including:

   a   Daily routine 'sweeps'.
   b   Communication functions in the case of referral/escalation.
   c   Co-ordination with other functions such as GP/community teams.

Many patients receive care every day that fits within the imagined alignment between care needs and specialist/non-specialist input. They will be assessed, have routine management taken care of on the ward, be seen by a nurse educator before discharge, and have home/after-care co-ordinated with multi-agencies if necessary. However, as with any health care microsystem, some threats to the desirable outcomes do emerge; in the following we outline how misalignments between demand and capacity threaten the quality of care.

Misalignments between demand and capacity

One of the problems in assumed ability to care, codified in procedure or policy, is that it does not tend to build in contingency for shift work, and the multiple people involved. Protocols focus on functions such as initial specialist assessment, ongoing glucose/nutrition management and elective specialist referral as if these are carried out by two omnipresent individuals: *the* specialist and *the* non-specialist. General principles of handover and team working may be part of health care training but the management of care by the wider team distributed throughout the hospital is achieved through adjustments that need to be learned and that require articulation and monitoring. Frequently, it was reported that information flow across the different professionals involved meant that not everyone understood the plan at all times:

We already are a multidisciplinary team [we have] dedicated psychiatrist [. . .] clinical psychologist and specialist nurses [. . .] there at the outset. One of the things that makes it more challenging is continuity of care because inevitably as an in-patient there's going to be multiple different people involved in the management with shift cover and making sure that people understand the issues and [. . .] to make sure there is a clear message where everybody [. . .] understands the plan, this is a very important aspect of treating them, it's one of the barriers. The other barriers are to make sure that the ward staff understand the issues [. . .] having staff understanding behaviour issues in patients and how to handle them. (Consultant, diabetes)

Such issues are magnified by a range of small but important misalignments such as non-standard insulin forms: 'Interestingly, if you go across different NHS hospitals, people have different forms they use.' But, perhaps the most important misalignment that has emerged is that the specialist team has ceased to be called upon only for difficult cases, and provides oversight and input into routine cases that would previously have been managed by ward staff alone. The emergent role of the specialist team illustrates the dynamic nature of the work system and shows how extra capacity can have the paradoxical effect of reducing capacity in other parts of the system:

The potential downside of having a good specialist team is that you actually de-skill the wards and there's a degree of learned helplessness in some areas where there's a lot of diabetes input where the ward staff do not engage because they leave it to the diabetes team and that is a negative thing in itself. (Consultant, diabetes)

Yeah, [. . .] ward nurses [. . .] I think they also feel like 'well I want to stick to what I know because I'm in a safe place', if I know any more then I will get more responsibility and I might get landed with all the patients with diabetes, or all the difficult ones so it's, it might be a safe place not to have more knowledge [. . .] I'm just thinking back to when I was a nurse on the ward, and the bit that I used to find difficult about diabetes was that I had some knowledge, but I didn't have enough [...] I think is what happens because the less you know, the more scared you feel [...] they have some knowledge but they don't feel confident to make the decision that need to be made (Diabetes Specialist Nurse)

These effects were relatively invisible for a time period before reaching criticality. The *short-term effect of introducing a specialist team* is that patients probably receive better than usual care, because the specialists' capacity to provide care a) exceeds the demands that routine cases present and b) still meets the demand in complex cases as before.

One ward consultant described how, citing 'not policy, *but practice*' he prefers to call the team for routine cases early on, and again before discharge:

So, there's the bit about getting the diabetes team involved early. [...] I think it's not policy but it's practice in the sense that because, they know that's how I work. They learn that when I go and see the patient and we say, okay, so the blood sugar's controlled now, what are you going to do next? Well, we can discharge him. And then we talk about, if we discharge him what will prevent this patient coming back again? [...] It's not hospital policy I guess. (Consultant, Non Specialist)

As the now very busy specialist teams cover complex and routine patients, and try to assist with pressured discharge planning in particular, the capacity of ward staff to autonomously cover routine care is already being eroded due to reliance on expert input. At this point the system has a chance to self-regulate or adjust. If a feedback loop existed to a) monitor the activity being carried out by various groups and b) recalibrate the specialist function back to higher level problem solving, then the capacity of ward staff could be re-established, and the specialists again used proactively on arrival and opportunistically when certain triggers appeared: However, in this case the specialist team were often seen as further stretched, to the point where they were unable to cover the large volume of routine cases being referred to them. At this point it is possible that outcomes could become worse if specialist cover is not available and ward staff expertise has been eroded:

They now rely on us to go and review their patients and I think we may have taken away that learning curve from them unfortunately; One of the potential drawbacks is that [...] well you could suggest that there be a de-skilling of others to manage inpatient diabetes. (Diabetes Specialist Nurse)

One nurse on the ward felt this was now the norm, calling it a culture:

So I think part of that though is the culture of not really recognising that this is something we should be doing as part of our routine care. Yes, we'll all look at pulse, blood pressure, temperature, but, oh yeah, they're diabetic, I forgot about that. (Ward Nurse)

A specialist nurse was worried about the future in this regard, in the context of financial cutbacks to service (see discussion):

It's very frustrating when we have enough people who we really do need to see. [...] And, of course, ideally we would see everybody but as you know with the explosion of diabetes it's and I think it's very worrying for the future. [...] Ten years down the line how are we going to be managing with all the cutbacks? (Diabetes Specialist Nurse)

Adjustments and 'work as done'

Here ward staff have adjusted to an opportunity presented by specialist cover, and thereby increased demand on the specialist team. Specialists have been able to cope to an extent, but then at a certain level a risk emerges as both specialist (overload) and non-specialist (under-use) capacities are eroded. By the end of the research period, this concern was explicit and further adjustments had begun to emerge in response to its associated threats. Specialist nurses reported increasingly trying to educate ward staff and attempting to avoid going to see patients:

I mean I think the specialist nurses are seen as people who should be called whenever you mention the word diabetes. [...] It is a big problem because 20 per cent of our in-patient beds are taken up with patients with diabetes and I think educating in-patient staff, the nurses about diabetes is a huge task and we are just chipping away [...] but it's like being at the bottom of a mountain [...] for us sometimes it is a bit depressing really in that people, or the nurses don't seem to think, don't think it through, even the simplest of things – high blood sugar – what do we need to do with their high blood sugar? It's not as though these issues, I'm sure aren't covered in training and so on but they call us – pick up the phone and if we say we're not going to see them; 'why are you not going to see that patient?' We try and educate the nurses even over the phone. (Diabetes Specialist Nurse)

Staff turnover meant education was an ongoing process: *'and then you win on some wards and then the staff change and you're back to square one'*. Many further adjustments were identified in the dataset: specialist staff often had to act based on unclear information (e.g., from telephone calls), leading to treatment based on a best guess as to what is required – what they referred to as a 'reasonable starting point'.

Specialist routinely had to make multiple calls to a range of people such as general practitioners, district nurses and community diabetes nurses to build an accurate picture of a case. There were many instances of trade-offs between 'ideal' and 'safe' glycaemic control, and adjustments to what was considered acceptable as an outcome, based on complex and dynamic treatment–need relationships, for example where patients would take extra insulin, or omit doses, due to cognitive impairment.

Finally, the blood sugar outcomes have to be assessed in the context of organisational outcomes. A telling quote from a senior Specialist Nurse was that the culture on assessment units has become *'get them seen and get them out'* with a focus almost entirely on hospital admission avoidance and/or length of stay, with associated pressure 'from above' on everyone to reduce both. The diabetes team are now seeing pre-admission patients in the Emergency Department, Clinical Decision Unit, Medical Assessment Unit and Acute Hub as well as other non-diabetes clinics *'before [they] even get to the wards'*.

Suggested interventions to improve care

Interviewees were asked for their perceptions of how to improve care. Some specialists in particular suggested decision-support tools to improve communication. The specialist function was seen as important mainly for discharging diabetic patients in a timely manner. This was felt to be sub-optimal and could be improved by seeing patients in order of complexity:

> *From our point of view, the need for the diabetic teams is more for helping us discharging patients. Admission is not a problem. [. . .] So, we're not prioritising our work properly. We prioritise it to say, we'll see all the DKA patients first then we'll see the routine patients, and last will be the education patients because they're going home anyway. Some system like that would be quite good. (Ward Nurse)*

Interestingly, suggestions often took the form of formalising some of the adjustments already being practised informally. So education and support for ward nurses, provided in response to the perceived deskilling, lead to a suggestion for a more formal training programme for junior and newly rotated staff. The gap between the specialist and non-specialist roles was seen as something that could be addressed by a link nurse or 'champion' on the ward. Again, someone had tended to function *de facto* in this specialist liaison role and the suggestion emerged from the perceived benefits. One specialist went a little further:

> *I think ideally we would have a member of the diabetes team affiliated to each ward, so we can go in and pick out who is a problem*

proactively rather than waiting to be 'bleeped' about a problem. (Specialist Registrar)

A final issue to note is that the recommendations made *rarely took the form of standardising process*. Looking from the outside, it might appear that, for example, having a single NHS-wide insulin pump, or recording form, might be an obvious way to align capacity with demand, but people seemed somewhat resigned to these variabilities, and suggested ways to work around them, rather than assume that the demand created by multiple devices and artefacts could ever be reduced.

Discussion and conclusions

In this final section we can discuss briefly the results of the study in terms of RE and the CARE model.

Exceptions, rules and the routine

Many clinicians were clear that there were no routine cases *per se*, meaning that complexity was more or less the norm in this work domain. The interviews showed the 'habitual solutions to the myriad of small problems that constantly stand in our way' (Wears, Hollnagel and Braithwaite, 2015b). This is the essence of the model, that misalignments are multiple and common, and adjustment is the norm.

The inpatient care cycle is often one of trialled action, variable success and variable feedback, leading to further action. Adjustment to treatment needs is of course part of everyday clinical work, but this can be contrasted with adjustments due to, for example, informational difficulties, which require many examples of 'chasing up' informal feedback to attempt to adjust to fluid situations. Seddon (2005) thus distinguishes between 'value demand' (work that is inherent to the aims of the organisation) and 'failure demand' (created by unwanted pressures and problems).

Workarounds and outcome trade-offs, for example, were very common and had become routine rather than exceptional (Debono and Braithwaite, 2015). This aspect of resilient health care, outlined clearly in previous volumes, means it can actually be hard to get people to articulate 'work as imagined' or the ideal/supposed pathways that once existed in the design process. People cannot recall easily how diabetes specialists and non-specialists might work in a difficulty-free scenario, given their rich experience in how things actually get done.

In discussing the results, clinicians reported that adjustment and trade-offs will only become more necessary and common, as the clinical pressure has increased due to: more patients admitted; complexity of cases has increased in last 5 years; the time to turn around problems reduced; diabetes management increasingly takes place in ambulatory settings as admission avoidance is the 'cultural zeitgeist'; and the 'environment of austerity' means 'rolling back' what can be offered given fewer Diabetes Specialist Nurses (known to be associated with less effective care; Bates et al., 2005).

Clinical Leads also talk about how high-level strategy has knock on effects over time, citing the example of a strategic focus on trauma, and the creation of an Acute Care Hub (ACH), having implications for how other specialities work.

The principal role: adjusting to variability

Much RE focus in health care (and indeed the wider focus of resilience engineering across many domains) is about variability. The CARE model characterises everyday work as coping with variable demands and capacities through adjustment. Many systems are organised formally to 'flex' in some way to cope with demand. In this case study, capacity is provided in the form of specialist and non-specialist staff, designed to cope with complex cases and conditions. However, there are many other adjustments that are not built in to the system in any formal way, but which emerge as important checks, balances and feedback loops. These can be at the level of individual actions, or co-ordination activities at higher levels, often between multidisciplinary team members.

The introduction of a specialist team, like any attempt to improve care, does not relate in a linear fashion to desired outcomes. Here, we observed the effect of initial improvement, followed by both expert complex care, but also a stretched specialist service and eroding ward level capability/ motivation. Recommendations centred on distributing expertise at the ward level (e.g., through ward-affiliated specialists) and improving ways of monitoring.

Efforts are currently being focused on tools showing a set of possible system states that can be used for anticipation; staff want to be able to track the system state over time, *before* re-attendance shows a particular discharged patient has fallen through the gaps, or incident/audit data show staff on the wards had become less capable.

Conclusions

In this case study we used the CARE model (see Chapter 4) to shed light on a system of diabetes care, how it was stressed or tested, and how resilience was maintained through adjustments, or broke down. In line with convention we elicited relatively unstructured accounts of events, and followed up areas of interest as they arose. However, we adapted the interview method to focus on the routine as much as the critical or exceptional. This re-orientation towards the routine and away from exceptional incidents is vitally important for examination of patterns and regularities (Hollnagel, 2015b).

We found that, as is common in health care, the system was set up to meet an anticipated demand, by placing resources where they would be most effective. The complexity of clinical care benefited from having specialist input, and in particular the ability of specialists to adjust to uncertain and dynamic conditions, and make priority decisions to ensure care was as safe as possible. However, the dynamics between specialist and non-specialist staff meant that some ability to provide care in ward staff was eroded, and this in turn required further adjustments on the part of the specialist team.

Chapter 14

Where process improvement meets resilience

A study of the preparation and administration of drugs in a surgical inpatient unit

Tarcisio Abreu Saurin, Caroline Brum Rosso,
Ghignatti da Costa Diovane and Simone Silveira Pasin

Introduction

In health care systems, resilient performance is not an option if care is to be delivered for patients. The nature of those systems implies features of complexity are ubiquitous, such as uncertainty and dynamic interdependences between functions (Fairbanks et al., 2014). No less true is the fact that efficiency pressures on health care systems are increasing. These pressures arise from a number of agents (governments, clients, regulators, among others), which are themselves subjected to similar pressures (Kenney, 2011). Therefore, a number of health care organisations have adopted process improvement philosophies originated from other sectors, such as lean production, which originated from the automotive industry. Well-known lean health care experiences, such as in the Virginia Mason hospital, started as a response to being 'better, faster, and more affordable' (Kenney, 2011, p. 129).

Given this context, process improvement projects are more and more a fact of life in health care, and thus these and resilience inevitably will meet each other at the frontline. Rather than dismissing such initiatives as fundamentally detrimental to resilience, we propose it is necessary to devise means for analysing how they influence (and can be influenced by) resilience. In fact, some earlier studies have already pointed out mutual learning opportunities between resilience engineering (RE) and lean (Saurin et al., 2013; Saurin et al., 2016), as well as some risks of interventions based on linear thinking and production pressures in complex systems (Woods, 2006). This chapter reports a practical experience in which the preparation and administration of drugs (PAD) in a University hospital was re-designed using traditional lean and process improvement tools, and then evaluated using insights from RE. The hospital, which is located in Southern Brazil, has 843 beds, about 5,000 employees (2,200 are nurses and technicians), and most patients come from the national public health care system.

Rationale for the research design

Organisational support for this study resulted from the fact that it was conducted in the context of a specialisation course (360 hours, spread over 1.5 years) in industrial engineering, given by faculty from the same University linked to the hospital. The hospital's top management made the decision to offer this course as a way of developing internal capabilities for improving performance. Thus, 30 employees of the hospital, chosen by top management, attended the course. Each employee started the course with a pre-defined assignment for applying the knowledge obtained from it. As part of this institutional initiative, a research project encompassing all sub-projects carried out by professionals/researchers was submitted and approved by the ethical committee. Thus, two authors of this chapter (DGC and SP) were nurses who attended the specialisation course, and they had a focus on PAD. This process was relevant due to: (i) its patient safety implications (Elliot and Liu, 2010) – 639 safety reports involving medications were made over 2014 and, from these, 378 involved the preparation and administration stages; (ii) the high frequency it occurs – about 4 million medications were dispensed over 2014; and (iii) the high volume of financial resources involved in the purchase of drugs.

Criteria were also adopted for selecting an inpatient unit as a pilot for assessing and re-designing the PAD process, as follows: (i) similarity with other units, which could make it easier for the replication of findings; (ii) use of recently acquired new technologies (e.g., automated dispensing cabinets, ADCs) and diversity of drugs, which were taken as proxy measures of the unit's complexity; and (iii) number of safety reports; a high number of reports was regarded as a proxy measure of the staff commitment to safety. Two of the researchers (DGC and SP) assigned scores to all inpatient hospital units (the higher the unit's relation to the criterion, the higher the score). As it turned out, the selected unit had 45 beds, dedicated to surgical patients. The unit's staff was formed of 14 nurses and 48 technicians, distributed over 5 shifts. This unit had twins ADCs, located side-by-side.

Design Science Research (DSR) was the methodological approach adopted. It is a way of producing scientific knowledge that involves the development of an innovative artefact to solve a practical problem, and simultaneously makes a kind of prescriptive scientific contribution (Holmstrom, Ketokivi and Hameri, 2009). As such, we were concerned with the development of an artefact, namely an intervention framework that integrated principles of RE and process improvement/lean. In fact, the concept of *systemic* intervention was adopted, which refutes the notion of flawlessly preplanned change based on accurate predictions of the consequences of the action (Midgley, 2003, p. 88). The researchers had an awareness of the complexity and its philosophical implications from the outset of the study.

As designers of the intervention, the researchers clearly had in mind the need for involving employees from the unit across all stages of data collection and analysis. The benefits and downsides of participation have been widely discussed by the human factors community, and it is agreed that participatory interventions usually produce better outcomes (Hendrick and Kleiner, 2001). Therefore, all problems and their prioritisation were either directly indicated by staff or validated with them after identification by the researchers.

Another important aspect of the research design relates to the composition of the design team (i.e., the authors of this chapter). We agree with Cook and Ekstedt (2016) in that the question of who are the 'engineers' in charge of engineering resilience should not be taken for granted, since well-intentioned but misguided engineers could cause more harm than good. In this respect, two of the researchers (TAS and CBR) were academics who had expertise in both RE and lean, but were not health care experts. The other two researchers (DGC and SP) were health care practitioners (20 and 25 years of experience, respectively, as employees of the hospital) who had recently been introduced to both subjects over the previously mentioned specialisation course, which included a 15-hour class on RE and a 30-hour class on lean.

This mixed composition proved useful to the extent that it helped to prevent 'solutions' based on linear thinking, in terms of clear-cut actions that should lead to expected results. Over the study it was noticed that wider systemic relationships of possible solutions tended to be easily missed in different degrees by all the researchers, who came to realise that systems thinking was not their strong point. Furthermore, a lesson learned from this arrangement of the research team relates to the value of systems-oriented mentorship when engineering resilience. In a sense, mentors can be interpreted as a type of redundancy, which is in itself a strategy for promoting resilience.

How data sources were selected

Data sources were selected in order to account for: (i) multiple sources of evidence; (ii) a mix of quantitative and qualitative data; (iii) perspectives of different agents involved in the socio-technical system; (iv) understanding of work-as-done; and (v) the description of the system according to both lean production and resilience engineering perspectives, using the value stream mapping (VSM) and the Functional Resonance Analysis Method (FRAM) as tools representative from each perspective, respectively. Two of the authors had previously applied both tools in an earlier study at the emergency department of the same hospital (Saurin et al., 2016).

On the one hand the VSM provides quantitative data, such as lead times and amount of work-in-process, in addition to focusing on client requirements – these set a reference for re-design efforts. On the other hand, the FRAM provides a much richer picture of work-as-done and therefore it puts quantitative data from VSM in a broader context, helping to understand how variability occurs and how it can propagate throughout the system. Overall, the selected data sources accounted for good practices of qualitative research (Flyvbjerg, 2011) and investigation of everyday clinical work (Hollnagel, 2015b), providing credibility for the findings.

How data were collected

Data collection occurred over 3 months (see Table 14.1), and it was concluded by the discussion of data analysis and interpretation with employees – i.e., the second focus group mentioned in Table 14.1. Over the data collection period, and based on the data collected, the VSM was applied following the steps proposed by Rother and Shook (1998): (i) to define a family of products or services to be mapped (in this case, the process of PAD); (ii) to design the map of the current state; (iii) to design the map of the future/desired state; and (iv) to develop an action plan to implement the future state. From the same

Table 14.1 Data sources and procedures for collecting data

Data sources	How data were collected
Participant observation	Insights from participant observation (15 hours) were recorded in a diary, emphasising the understanding of work constraints and the flow of drugs, information and people. Observations were often associated with informal conversations with employees, in order to clarify technical issues and routines.
Time and motion studies	Based on observations and filming, the sequence of operations involved in the PAD was mapped – estimates of walking distances were also made – and the time of each operation was recorded. There were four basic types of operations: processing, waiting, transporting and inspecting. From the lean view, just processing (e.g., administering drugs) adds value to the client, and so the others should be minimised as much as possible.
Documents	Consultation of records of incident notifications related to PAD, standard operating procedures (SOPs) related to PAD, medical pre-scriptions, patient charts, evaluation reports filled out by patients or family members at discharge and blueprints of the unit.
Focus group	Two meetings (2.5 hours each) organised as focus groups were held. Each meeting was attended by 13 employees – not all had the same job. The first meeting focused on the identification of the clients of the PAD, their requirements and problems. The second meeting involved refining the lists of requirements and problems.

database, and based on Hollnagel (2012), the FRAM was applied by: (i) defining the purpose of the FRAM analysis (in this case, to model everyday work in order to assess the impact of re-design measures); (ii) identifying and describing the aspects (i.e., input, output, time, control, preconditions and resources) of the functions, which refer to the acts or activities that are needed to produce a certain result; and (iii) identifying the real variability of the outputs of each function, both in terms of precision and time. Through these steps, a FRAM model of everyday work of the PAD process was developed using the software FRAM Model Visualizer 2.0. For a detailed description of both the VSM and FRAM we refer the reader to seminal books by Rother and Shook (1998) and Hollnagel (2012), respectively.

How data were analysed

Content analysis of notes from meetings, observation diaries and other documents was the main data analysis method. As such, researchers looked for excerpts of text that showed evidence of: (i) stages (for the VSM) or functions (for the FRAM) of the PAD process; (ii) clients of PAD; (iii) clients' requirements; (iv) problems according to the process stages or functions in which they occurred – this also offered insight into the variability of the outputs of the functions, information needed for applying the FRAM; and (v) potential solutions for the problems. Data from time and motion studies were organised in tables and graphs in order to illustrate the extent of some problems. For instance, a 'spaghetti' diagram (Chiarini, 2013) was drawn to show the confusing and lengthy workflow of nurse technicians during the PAD.

How the analysis results were interpreted and gave rise to recommendations

As a starting point, each problem was analysed from the perspective of six guidelines for the management of complex socio-technical systems, which are in line with RE and were developed by Saurin et al. (2013) and Righi and Saurin (2015): (i) give visibility to processes and outcomes; (ii) anticipate and monitor the impact of small changes; (iii) encourage diversity of perspectives when making decisions; (iv) design slack; (v) monitor and understand the gap between prescription and practice; and (vi) create an environment that supports resilience. A meta-guideline (Clegg, 2000) states that the other guidelines are contingent, which means that they can be counter-productive under certain circumstances. Based on insights from this analysis, supported by the FRAM model of everyday work, recommendations for system re-design were made and once again checked against the six guidelines.

Results

Overview of the VSM and FRAM results

From the VSM, 19 process stages and 6 clients were identified: patients, family members, nurses, nurse technicians, society and physicians. These last ones were both suppliers and clients, as they provided prescriptions that started the process of PAD and used the records of administration for preparing a new round of prescriptions. These clients had 41 requirements associated with 45 problems, which were defined as recurrent situations that compromised one or more requirements. From the FRAM model representing everyday work, 16 functions were identified. Based on the data collected, the researchers considered that the outputs of 12 of these functions had substantial real variability in terms of precision and/or time. While there was a clear association between the problems identified from the VSM and the variability identified from the FRAM, the latter offered a more nuanced perspective of the origins of shortcomings. For instance, while the key process stage or function in both models (VSM and FRAM) is the administration of drugs, the VSM conveys a linear view in which this is directly preceded only by one previous stage (i.e., identification of patient at the bedside) and directly followed by just one stage (i.e., discard residuals). By contrast, the FRAM indicates that <administer drugs> directly interacts with at least six other functions.

Analysing a problem from the perspective of complexity and RE

We selected one problem in order to illustrate the analytical approach used. It refers to the fact that medications stored in the ADC at the unit are not received from the pharmacy labelled with the name of the corresponding patient. This identification needs to be made at the unit by a nurse technician. They have to retrieve the drug from the ADC, pick up a tag with the patient name from a folder available at the nursing station, and manually write down: date, drug name, administration time, dose, via and signature of the technician. Some of this information involves the use of acronyms, which have dozens of types and sometimes the same acronym has completely different meanings. After writing down information on the tag, it needs to be attached to the drug. Thus, if a patient needs several drugs at a certain time, several tags will be necessary; one attached to each drug. Drugs for all patients under the care of each technician are located on trays in which tags may be easily lost. A few technicians prefer to prepare and administer drugs on a one-patient basis – i.e., a tray only contains drugs to be given to a certain patient at a certain time. It is worth noting that only the patient name is printed in the aforementioned tags in the folder; the secretary of the unit does this every day by evening. The secretary puts in the folder one page of unfilled tags for each patient, although sometimes a larger number is necessary and the technician needs to print additional tags. Overall, the described activities add to the workload of technicians, creating competition for shared resources such as printers, tags and work space. Furthermore, professionals who attended the focus groups considered the task of associating drugs with patients as error-prone, affecting patient requirements related to administration of the right drug at the right time. Table 14.2 presents the analysis of this problem according to the six guidelines previously mentioned.

In addition to the conclusions arising from Table 14.2, a broader insight from the analysis refers to the realisation that not all levels of the system adjust effectively over time. This creates brittleness at the interface between the well adapted and the not adapted sub-system. An example is related to the use of very expensive new medication. This results from adaptive behaviour by higher hierarchical ranks, which may be aware of more effective treatment and manage to purchase it. However, such new and expensive medication can be placed into an inefficient and risky process of PAD, which may not make effective use of the new treatment – i.e., the PAD process does not adapt as effectively as the processes of selecting and purchasing drugs. In the words of an experienced nurse who worked at the unit, *'over the last 30 years a number of technological changes have been made, but my work did not change anything, I have just coped with the changes'*. This may be interpreted as a complaint for overusing reactive resilience at the frontline, as a result of a brittle interface with resilient actions by higher hierarchical ranks. This example also suggests that having resilient parts is not sufficient for having a resilient system.

Analysing countermeasures from the perspective of complexity and RE

Researchers have tentatively considered countermeasures to tackle the problems cited in Table 14.2. RE, in this context, works as a quality check of these countermeasures, by: (i) assessing whether they are in line with the six guidelines; and (ii) assessing how their impact may propagate throughout functions. Three countermeasures under discussion by researchers and hospital staff were selected to illustrate these points (Table 14.3): the delivery of drugs from the pharmacy with tags identifying patients; to define the window of time for retrieving drugs from the ADC based on explicit and consensual criteria; and the use of a help chain. As to this last measure, regardless of the fact it is often used in hospitals, it is not usually known by that name. The term 'help chain' is normally used in lean production environments and, based on Spear and Bowen (1999), it is defined here as follows: a standardised routine for the identification and solution of abnormalities, which is triggered by a visual device (*andon*, in

Table 14.2 Analysis of the selected problem according to the guidelines for managing complex socio-technical systems

Guidelines	Current state
(1) Give visibility to processes and outcomes	Drugs in the ADC have no tags associating them with patients.
(2) Monitor and understand the gap between prescription and practice	There is an incident reporting system, and employees from other units carry out audits of standardised operating procedures (SOPs). However, results from audits are not widely disseminated and these usually do not give rise to changes.
(3) Encourage diversity of perspectives when making decisions	PAD is an individual task. However, there is no standard protocol regarding how technicians should proceed in order to request help in case of doubt (e.g., unclear prescription, malfunction of the ADC).
(4) Design slack	Slack seemed to be mostly opportunistic. For instance, in the case of damaging a medication, the technician needs to go to the pharmacy, and present a written justification for a new dose. As this can be time consuming, workarounds can occur – e.g., the technician can remove a drug while another worker keeps the ADC opened. An example of designed slack refers to the ADC programme which allows the retrieval of drugs over a window of time, from 2 hours before to 1 hour after the scheduled administration time. If this window is missed, technicians can retrieve drugs from the ADC by changing the administration schedule. This change may be time consuming.
(5) Anticipate and monitor the impact of small changes	There is vulnerability if the physician changes the scheduled time – e.g., the technician retrieved the drug from the ADC and prepared it at 3 pm (but it was due at 4 pm), while the physician changed (through the computerised system) the prescription at 3.05 pm, replacing the prescribed drug with another one. Such change can have a large impact on the health outcomes of the patient.
(6) Create an environment that supports resilience	Training is focused on rule-following and there are no mechanisms for identifying and disseminating good practices that fill the gaps in SOPs. Employees complained of time pressure during handovers. Thus, problems such as delays in administering drugs are unlikely to be informed to co-workers, hindering performance adjustment.

Table 14.3 Analysis of the impact of countermeasures

Countermeasure	Guidelines affected	Propagation throughout functions
To implement a help chain	This practice: (i) provides visibility to abnormalities; (ii) encourages team decision-making; (iii) offers an opportunity to check prescription against practice; (iv) provides cognitive slack for solving problems, while at the same time removing slack from functions undertaken by helpers; and (v) createsconditions for successful adaptations of rules.	If a help chain is used, technicians would no longer need to leave the ward or workstation in order to look for a co-worker capable of helping. As a drawback, functions carried out by helpers could be interrupted.
To receive drugs labelled with patient's name from the pharmacy	Essentially, this measure would increase process visibility. It may also create slack, since technicians will be freed up of filling tags.	Function <fill out tags> would be less variable since it would be carried out in a more controlled environment (i.e., the pharmacy). Due to a more stable output, functions that use it as an input would be more stable as well.
To define the window of time for retrieving drugs from the ADC based on explicit and consensual criteria	Slack would be created if the window was either extended or abolished. By contrast, slack would be reduced if the window was shortened. The ISMP (2010) reports undesired side-effects created by a policy of enforcing a '30 minutes-rule', which implies the administration of drugs within 30 minutes before or after the scheduled time. Of course, increased slack could have unexpected side-effects too.	Shortening the window of time would very likely increase the variability of other functions, since more time pressure would be put on workers. The effects of extending or abolishing the window of time would not be as clear. In an optimistic view, it could reduce workarounds to get drugs from the ADC when the window is missed.

lean jargon) to request help from support areas, which in turn should go to the requested location, jointly discuss a solution and produce organisational learning.

Analysis of propagation throughout functions was made with the support of the FRAM model of the everyday process of PAD (Figure 14.1). From this, it can be noticed, for instance, that the help chain could be part of the control aspect of some other functions, in which the need for asking for help from co-workers is more likely to occur. Similarly, once drugs are received from the pharmacy labelled with the patient's name and treatment information, the variability of the output of the function <fill out tags> is likely to be reduced. Concerning changes in the ADC window of time for retrieving drugs, this would affect the control aspect of <retrieve drugs from the ADC>.

Conclusions

Health care systems are constantly evolving, and an increasing portion of these dynamics arises from intentional and formal process improvement initiatives. Health care practitioners and academics should face these initiatives as a lever for introducing RE principles. Initially, as reported in this chapter, the use of RE can take the form of assessing and refining 'solutions' originally devised through process improvement projects. As shown in this chapter, RE helps to explain *why* a certain countermeasure makes sense (or not) in a complex environment, and it also has a prophylactic role in terms of making it clear that undesired side-effects of 'solutions' need to be explicitly recognised. For example, it was shown that the underlying principles of a help chain are aligned to RE, which can contribute to reducing implementation resistance in health care. By contrast, new social interactions are triggered by the help chain, with effects that cannot be completely anticipated.

Over time RE must be an integral element of process design, right from the beginning. In this sense, the reported approach for introducing RE in practice was modest but fitted to the context in which it was used: a large health care public organisation that provides care in a number of different medical specialties, and in which practitioners were not familiarised with this paradigm. In such a highly complex setting, a quick and pervasive introduction of RE would be both unrealistic and inadequate.

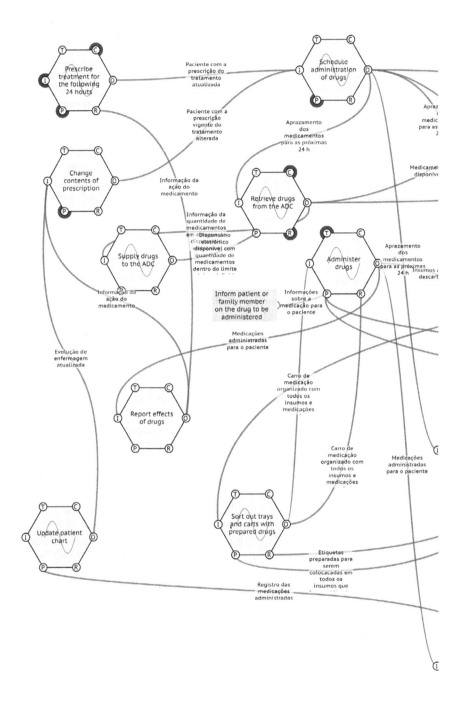

Figure 14.1 FRAM model of the everyday process of PAD.

Notes: (i) highlighted lines represent couplings between <fill out tags> and other functions;
     (ii) aspects of the functions are represented by I (Inputs), O (Outputs), T (Time),
     C (Controls), P (Preconditions), R (Resources); and (iii) re-design measures
     discussed in this section are listed besides the functions in which they can be used.

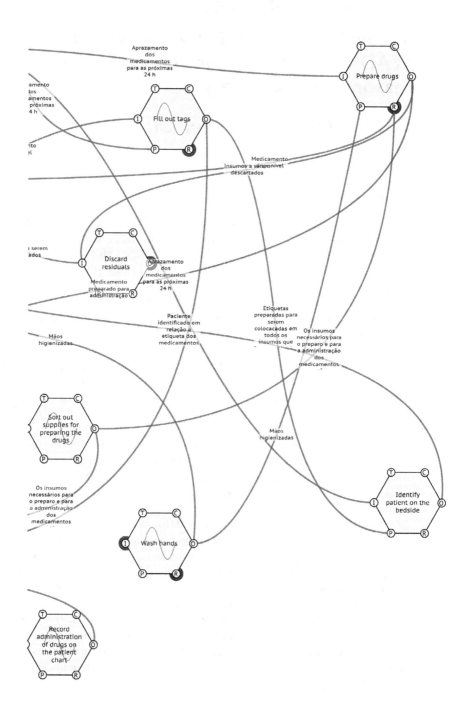

Aprazamento dos medicamentos para as próximas 24 h

amento los amentos próximas 4 h

nto l

Fill out tags

serem ados

Discard residuals

Medicamento Insumos a serem descartados

Medicamento preparado para administração

Aprazamento dos medicamentos para as próximas 24 h

Prepare drugs

Paciente identificado em relação à etiqueta dos medicamentos

Etiquetas preparadas para serem colocadas em todos os insumos que

Os insumos necessários para o preparo e para a administração dos medicamentos

Mãos higienizadas

Sort out supplies for preparing the drugs

Mãos higienizadas

Identify patient on the bedside

Os insumos necessários para o preparo e para a administração dos medicamentos

Wash hands

Record administration of drugs on the patient chart

Chapter 15

The Safety-II case

Reconciling the gap between WAI and WAD through structured dialogue and reasoning about safety

Mark A. Sujan and Peter Spurgeon

Introduction

The way everyday clinical work unfolds (work-as-done/WAD) is usually different from the way managers and administrators think about clinical practice (work-as-imagined/WAI) (Hollnagel, 2015c; Hollnagel, 2016). For health care systems to become more resilient, it is important that these two perspectives are better reconciled (Braithwaite, Wears and Hollnagel, 2017). Health care professionals working at the sharp end have to make adaptations and trade-offs in order to provide safe and good quality care. When the gap that exists between WAI and WAD is unappreciated, there is a risk of almost a type of 'system complicity' among staff in not exposing issues that might trigger the need to redefine WAI (Sujan, Spurgeon and Cooke, 2015b). As a result, improvement efforts are likely to introduce additional constraints and burdens on clinical practice (Debono and Braithwaite, 2015). How can the reconciliation of WAI and WAD be achieved in practice?

In this chapter we describe an illustrative case study of the practical application of an approach that is intended to foster dialogue around safety among health care stakeholders. The approach, the Safety-II case, is adapted from the safety case used in safety-critical industries, and it aims to provide confidence that stakeholders are reasoning systematically about safety (Sujan et al., 2015a; Sujan et al., 2016). We argue that, appropriately adapted for use in health care, the safety case approach might contribute towards creating greater awareness of everyday clinical work among different groups of stakeholders. Clinicians, managers and others involved in improving clinical practice can use this approach to take stock of their safety position in an honest and transparent way, and open up their assumptions about safety to review and critique. In this way, the approach might serve as a bridging structure in the thinking about Safety-I and Safety-II, and stakeholders might be better able to align their patient safety improvement efforts with WAD, rather than base improvements simply on idealised assumptions of WAI.

The chapter is organised as follows: the next section gives a brief summary of safety cases as used in traditional safety-critical industries. Then, a critique of safety case practice from a Safety-II perspective is provided. This is followed by an illustrative case study of the development of a Safety-II case in the context of clinical handover in emergency care. We then discuss the potential role of the Safety-II case in health care more generally, and we conclude the paper with implications for policy and for practice.

The reasoning behind the safety case approach

In safety-critical industries, such as the oil and gas industry and the nuclear industry, there has been a shift in regulatory regimes over the past 20 years from prescriptive towards goal-based approaches. Under a prescriptive regulatory regime, the regulator specifies the technical and organisational provisions that need to be in place in order to operate a safety-critical system. This approach has proven to be ineffective due to the fast evolution of technology, and the complexity of modern systems (Hawkins et al., 2013). With the goal-based regulatory approach, the regulator specifies goals that need to be achieved, but leaves the specific ways in which the goals are met to the manufacturers and operators of systems (Sujan, Koornneef and Voges, 2007). This provides flexibility, but also shifts the responsibility of demonstration that a system is safe to the duty holders. In the UK and several other countries this demonstration is fulfilled through the development and submission of a safety case (Sujan et al., 2016).

The safety case provides structure to an organisation's safety manage-ment activities (Sujan et al., 2013). With the safety case an organisation argues that a system is acceptably safe to operate within a given context. The defining feature of a safety case is the presence of claims (e.g., that the system is acceptably safe), arguments and supporting evidence (Maguire, 2006). Without an explicit safety case, organisations often do not present an argument, and simply supply a wealth of unexplained (and potentially arbitrary) evidence; or they present arguments about safety without the actual evidence to substantiate these. By developing a safety case, organisa-tions can assure themselves, as well as the regulators and the public, that they have undertaken reasonable, systematic and rigorous efforts to build and to operate safe systems.

There are many different ways in which one might argue that a system is safe. Professionals will often provide a range of claims, arguments and evidence including references to well-motivated and qualified staff, the absence of recent adverse events, and the presence of continuous improve-ment programmes. All of these pieces of information bear relevance, yet none is sufficient by itself to create a comprehensive and defensible case. Typically, an industrial safety case consists of a risk-based argument and a confidence argument. The risk-based leg argues that all relevant risks have been identified and reduced to acceptable levels. The confidence leg argues that the evidence presented is sufficiently trustworthy (Hawkins et al., 2011).

While there has been critique of the safety case approach (Leveson, 2011), there are good reasons for exploring the adoption of this practice in health care (Sujan et al., 2016). Safety cases might support health care organisations in establishing proactive safety management practices, and they can create transparency and openness regarding assumptions about safety. However, the adoption of safety cases in a health care context is not straightforward due to the different level of maturity of safety management practices in health care, and the complexity of modern health care systems.

Safety cases from a Safety-II perspective

Current safety case practice is firmly rooted in traditional Safety-I thinking. The high-level claim of a safety case typically states that a system is acceptably safe to operate (with various qualifications). The notion of acceptability is then linked to the concept of risk. In the UK, a system is regarded as acceptably safe if risk has been reduced as low as reasonably practicable (ALARP) (Health and Safety Executive, 2001). Risk is commonly expressed as the combination of the likelihood of occurrence of a failure, and the severity of the resulting consequences. The safety case creates an argument and provides evidence that risks associated with hazards have been reduced ALARP.

From a Safety-II perspective, this kind of thinking creates both practical as well as conceptual problems for the application of safety cases in health care. At the practical level, there does not exist an agreed notion of acceptable levels of risk in health care (Sujan et al., 2016). Safety management in health care organisations is still predominantly reactive, and is frequently driven by externally set targets about reduction of the number of specific adverse events, such as pressure ulcers, patient falls or urinary tract infections. Where improvement teams adopt more proactive risk-based approaches, such as Failure Mode and Effects Analysis (FMEA) (Sujan and Felici, 2012; Apkon et al., 2004), there is a lack of guidance as to which risks should be addressed and how to determine the extent to which risks should be reduced. Practice is, therefore, very variable and is dependent on individual initiative (Sujan et al., 2015a).

At the conceptual level, problems arise from the focus on safety as the absence of incidents, accidents and disasters (Hollnagel, 2014b). Traditional safety cases, in essence, argue that the extraordinary catastrophe will not happen, but there is no account of the role of ordinary everyday work (Sujan, Pozzi and Valbonesi, 2016). There is the danger that the safety case is reduced to the description of a collection of well-intentioned barriers, defences and safeguards that ultimately represent instances of formal assumptions about how work should be carried out, i.e., WAI. Recent research in resilient health care (Braithwaite, Wears and Hollnagel, 2015; Fairbanks et al., 2014; Sujan, Spurgeon and Cooke, 2015a; Sujan et al., 2015c) as well as the chapters contained in the previous books in the present series (Braithwaite, Wears and Hollnagel, 2017; Hollnagel, Braithwaite and Wears, 2013c; Wears, Hollnagel and Braithwaite, 2015a) have demonstrated the importance of considering not only what can go wrong, but also studying and supporting what goes right.

Modern health care systems are characterised by changing demands and finite resources giving rise to competing organisational priorities, such as the management of patient flows and time-related performance targets (Sujan et al., 2015c; Anderson et al., 2016). Health care systems might be regarded more appropriately as Systems of Systems (Harvey and Stanton, 2014) or Complex Adaptive Systems (Braithwaite et al., 2013). The complexity of this context creates tensions (Sujan, Rizzo and Pasquini, 2002) that clinicians have to resolve and to translate into safe practices through dynamic trade-offs on a daily basis (Sujan, Spurgeon and Cooke, 2015a, 2015b). Managing tensions through trade-offs forms part of clinicians' everyday work, and they do not normally distinguish such activities from their technical work and the patient care that they undertake (Cook, Render and Woods, 2000; Hollnagel, 2009a). Approaches rooted in the Safety-I perspective run the risk of not properly appreciating the positive contribution of performance adjustments and dynamic trade-offs to the delivery of safe care (Cook, 2013; Hollnagel, 2009a; Hollnagel, 2014a).

These criticisms relate to the definition of safety as Safety-I rather than to the safety case concept as such. The safety case, as described above, is best understood as a communication tool that integrates claims about a system, and the corresponding arguments and evidence into a systematic structure. This provides the opportunity to explore the notion of a Safety-II case – a structured communication tool intended to foster dialogue among stakeholders about safety, which provides a bridging structure for Safety-I and Safety-II thinking, and thereby creates awareness of the gap between WAI and WAD.

Case Study: Safety of clinical handover in emergency care

Rationale

Improving the safety of clinical handover in emergency care has been recognised as a priority in health care systems around the world (Cheung et al., 2010; Sujan et al., 2015b). Handover is a frequent and highly critical task in clinical practice, as it ensures continuity of care, and provides clinicians with an opportunity to share information and plan patient care.

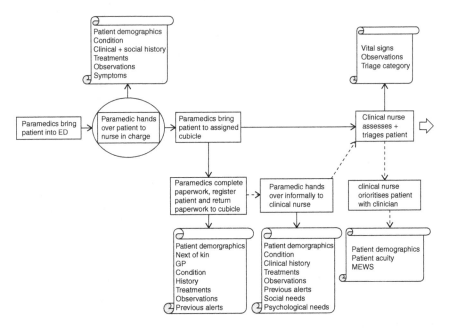

Figure 15.1 Emergency care pathway.

The case study is concerned with clinical handover in the emergency care pathway. In emergency care, handover is a particularly important activity, but it is also highly vulnerable due to high patient acuity and frequent situations of overcrowding in the emergency department. Within this pathway, handover occurs between members of ambulance crews and emergency department staff, and between emergency department staff and hospital ward staff. Figures 15.1 and 15.2 show process map representations of the pathway (Sujan, Spurgeon and Cooke, 2015a). Oval shapes indicate handovers with transfer of responsibility for patient care. Scrolls represent examples of information that is communicated during the handover.

Handover continues to be a stubborn problem (Sujan, Spurgeon and Cooke, 2015a). Frequently suggested improvement interventions aim to reduce handover and communication failures through standardisation of the handover conversation using a range of mnemonics, such as SBAR (Situation, Background, Assessment, Recommendation), but these have struggled to deliver the anticipated improvements (Sujan et al., 2015b). A possible reason for the perceived shortcomings of these improvement interventions might be that they are based on a too narrow Safety-I perspective. From this perspective, handover is regarded as a discrete activity, and handover failures constitute deviations from prescribed

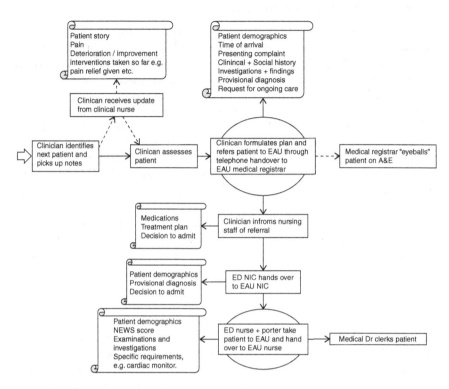

Figure 15.2 Emergency care pathway (continued).

practice (WAI). In actual practice, handover is a very complex activity, which may serve different and, sometimes, competing purposes, such as the management of capacity and demand, delegation of care, transfer of clinical, social and logistical information, and highlighting and prioritisation of specific aspects of patient care (Sujan et al., 2014). In this instance, understanding WAD can shed light on how clinicians usually embed handover in their activities to meet these different aims successfully, and why, at times, a particular handover turns out to be insufficient or inadequate to support the delivery of good quality care to a specific patient.

Therefore, the purpose of the development of the Safety-II case for this pathway is to provide assurance that stakeholders have really understood their current safety position. This is achieved by arguing that both what might fail as well as what goes right has been understood, i.e., a combination of the Safety-I perspective and the Safety-II perspective. In this way, both WAI and WAD are represented, and their integration is achieved by drawing upon shared evidence, such as process maps, observations and interviews.

Data sources and data collection

The data sources that can feed into a safety case are plenty, and they depend on the specific context. In this example, the following data sources were used: a half-day process mapping session with nine stakeholders from the ambulance service (two participants), the emergency department (five) and hospital wards (two), was held to map out the process at a high level taking into account the perspectives of different roles within the pathway; subsequently, two half-day FMEA sessions were carried out using the process map as the basis with the aim of identifying key risks; a half-day Functional Resonance Analysis Method (FRAM) session (Hollnagel, 2012) was held in order to augment the discussion about failure modes with discussions about performance variability (Sujan and Felici, 2012); observations were conducted in the emergency department for a period of 7 months (approximately 4–6 hours per week); and semi-structured interviews were undertaken with 15 participants (4 participants from the ambulance service, 8 from the emergency department, 3 from hospital wards).

Data analysis

The process maps and the FMEA sheets resulted in a systematic risk analysis. FMEA considers the possible failure modes of process steps, and prompts participants to reason in a structured way about (a) possible reasons for the failure, (b) possible consequences, (c) where possible, the likelihood of failure, and finally (d) potential mitigation strategies. FRAM was used to reason about and establish how the variability of a particular function can propagate throughout the pathway. Using FRAM, participants' reflections on how they deal with variability, and how this variability might affect patient safety, were combined with the FMEA risk analysis. Observations and interviews were analysed using Thematic Analysis. As part of the analysis tensions and trade-offs were documented.

Interpretation

A high-level safety argument is shown in Figure 15.3. The claim that the safety position has been understood is supported by an argument consisting of three legs: (i) a traditional Safety-I argumentation claiming and providing evidence that the current risk position has been understood; (ii) a Safety-II argumentation claiming that the role of performance variability (i.e., WAD) has been understood; (iii) and an argument that the evidence presented in the safety case is trustworthy.

The risk-based claim could be further broken down into claims that key risks have been identified and that steps have been taken to control these.

The evidence (not shown in Figure 15.3) to support these claims could come from the process mapping sessions, the FMEA and FRAM analyses, as well as subsequent options analysis, where risk controls are evaluated. Key risks that have been identified in the example result from inadequate patient flows and include: delays in ambulance crew handover, inadequate patient transfer handover owing to unfamiliarity with the patient, information loss resulting from different information needs of the handover parties, and refusal to accept patient referrals (Sujan et al., 2014). An example is provided in Table 15.1.

The Safety-II claim in the example has been broken down into two claims stating that tensions and trade-offs have been identified, and that steps have been taken to support the management of performance variability through strengthening of the abilities to anticipate, adapt, monitor and learn. The evidence (not shown in Figure 15.3) to support the former claim comes from the FRAM analysis, and from the observations and interviews, while the evidence to support the latter claim could come from an operationalisation of the Resilience Analysis Grid (Hollnagel, 2011a), which also serves as justification for the argument strategy. Various tensions and the resulting trade-offs that health care professionals make were identified in this case study (see the examples provided in Table 15.2). In all of these examples, staff utilise their subjective assessment of the demands of the current situation in order to dynamically resolve tensions through trade-offs.

Finally, the confidence claim refers back to the way the data collection and data analysis activities have been carried out. In this way, similar to the assessment of the quality of a scientific article, the reader of the safety case

Tabel 15.1 Example of a handover risk identified through risk analysis (Safety-I perspective)

Delay in handover from paramedic to nurse in charge.	In those cases where the ambulance crew are queuing and waiting to hand over a patient to the nurse in charge, there may be the possibility that the patient deteriorates while in the queue, potentially requiring more intensive treatment later on, for example the development of severe sepsis that results from delays in the recognition and treatment. Causes for this are ED overcrowding or observations that have not been re-checked in the queue, possibly because ambulance crews have handed over to another crew who are now looking after several patients. Delays are perceived to happen every day. Possible risk control interventions might include: introduction of a queue nurse, introduction of a rapid assessment triage team, and education to the public about appropriate use of the ED in order to reduce overcrowding.

can build an appreciation of how robust the evidence to support the claims is, and where the gaps in the evidence are.

Can the Safety-II case support reconciliation of WAI and WAD?

Safety management in health care is still largely reactive and centred around externally set targets. Such targets are usually based on something that can be conveniently counted, including both adverse events (e.g., patient falls) as well as the presence of barriers and safeguards (e.g., venous thromboembolism risk assessment completed). Improvement interventions that are commonly adopted focus frequently on the introduction of further barriers, standardisation of practice, and the provision of education to staff on the importance of complying with best practice. However, this approach is not usually based

Table 15.2 Examples of trade-offs staff make during handover in emergency care (Safety-II perspective)

Trade-off	Example
Ambulance crews trade off the unmet need in the community with the needs of the patient under their care.	When queuing the ambulance crew might decide to hand over to another crew waiting in line in order to get back into the community faster, but the second crew will not able to give as detailed a handover.
	After the first handover to the nurse in charge the ambulance crew might decide to wait in the emergency department and hand over a second time to give a detailed account of the patient's psycho-social needs, but this risks their violation of the turn-around target.
Emergency department staff trade off the needs of the patient they are referring to a ward with the resources available (beds) for other patients.	In situations of overcrowding, emergency department staff might decide to force referral of a patient by using certain keywords that make refusal of the referral more unlikely, but this might lead to patients being sent to the wrong place.
Ward staff trade off the needs of a patient being referred with the needs of the emergency department and their own availability of resources.	Ward staff try to ensure that patients are referred to the right place and that the ward has sufficient resources before accepting new patients. In situations of known emergency department overcrowding, ward staff might accept patients more readily even if subsequently they might have to refer the patient to another ward.

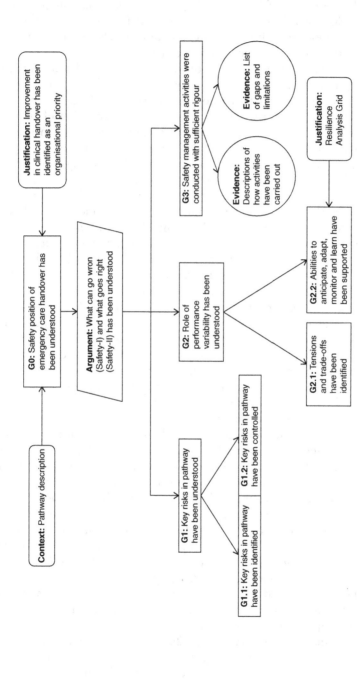

Figure 15.3 Safety-II case for emergency care handover – High-level argument.

on a thorough understanding of WAD, and it prevents a critical reflection about the adequacy of WAI. The safety case approach might support organisations in adopting a more structured way of understanding care pathways, and it might promote proactive reflection on both vulnerabilities and the contribution of performance variability and trade-offs.

This has been demonstrated in this chapter through the description of a clinical handover case study. The safety case approach prompted consideration of risks in a systematic fashion (Safety-I), and a large number of risks were identified. This proactive and systematic identification of risk already goes a good deal beyond what is currently common practice in health care safety management. In addition, the explicit consideration of what goes right, i.e., of the role of performance variability and dynamic trade-offs (Safety-II), provided a richer picture of WAD. Together, these insights can form the basis for improvement activities that might be better suited to account for the complexity of handover as a social and technical activity. This approach does not pit Safety-I against Safety-II, but aims to provide a bridging structure between the two ways of thinking about safety.

Clearly, the safety case approach needs to be adapted in order to be of use in health care settings. The high-level claim of systems being 'acceptably safe' will most likely not be helpful in this domain. Health care organisations and systems do not have a shared framework for determining acceptable levels of risk (Sujan et al., 2016). It might be argued that this presents an opportunity for the adoption of safety cases in health care rather than a problem: from a Safety-II perspective it makes more sense to reason about where an organisation is at in terms of safety, and how it navigates the complexity of competing organisational priorities.

Arguably, the biggest potential contribution of the Safety-II case is that it provides health care organisations with a structure and a framework to articulate and to reflect on their beliefs about safety. With the engagement of a wide range of stakeholder groups, beliefs and assumptions about clinical work (WAI) can be expressed explicitly and contrasted with the realities of how health care professionals manage competing priorities through dynamic trade-offs (WAD). Supported by the above handover safety case, it is arguably rather easy to spot where assumptions that accompany the introduction of standardised communication protocols will fall short of actual practice. For example, when ambulance crews are unable to hand over because they are stuck in a queue, standardisation of the communication will not provide much support!

While the risk-based argument leg (Safety-I) of the safety case is reasonably well understood in theory, the argument strand about performance variability requires further research. In the example, we have alluded to the Resilience Analysis Grid (Hollnagel, 2011a), which is an empirical set of questions that organisations might use to think about their ability to anticipate, to adapt, to monitor and to learn. In practice this requires further elaboration and operationalisation. Other argument strategies and sources of evidence might be considered. For example, the resilience model described in Anderson et al. (2016) could form the basis for a Safety-II argument. On a more empirical level, the 10-Cs model proposed in Johnson and Lane (2016) specifies resilient traits that can be taught and implemented in practice. Claims about the ability to learn could be supported by evidence drawn from approaches to organisational learning based on Safety-II concepts, such as the Proactive Risk Monitoring (PRIMO) tool (Sujan, 2012; Sujan et al., 2011) and other decentralised learning approaches aimed at complementing centralised incident reporting systems (Sujan, 2015; Sujan and Furniss, 2015).

Conclusion

Reconciling the gap that exists between the way that designers and managers think about clinical work and patient safety (WAI) and the way that health care professionals deliver safe and good quality care on a daily basis (WAD) is a prerequisite for resilient health care systems. This is, first and foremost, an exercise in communication and building trust between stakeholders. As a communication tool, the Safety-II case might support stakeholders in establishing a structured dialogue about safety in an open and transparent fashion. Assumptions about clinical work can be integrated with the realities of everyday clinical work in a constructive and mindful way.

At a practical level, health care organisations need to become more proactive and more mindful about the way they manage safety. Being more proactive requires taking greater responsibility for safety improvements, and relying less on externally set targets and drivers. Being more mindful implies that health care organisations need to have awareness and be supportive of the positive contribution of the trade-offs that health care professionals need to make on a daily basis.

While the Safety-II case might not, and probably should not, become a regulatory instrument, much might be gained by having regulatory buy-in and support. Health care organisations require assurance that they will not be penalised for exposing vulnerabilities in their system, and for investing resources in understanding and supporting performance variability rather than in meeting static performance-related targets.

Both health care service providers and regulatory bodies require education about the principles of resilient health care as well as about proactive safety management more generally. Stakeholders require access to practical examples and guidance that takes into accounts the specific characteristics of health care as an industry. In addition, there is an urgent need for further empirical evidence of the effectiveness of resilient health care approaches in practice.

Acknowledgements

This work was supported by research grants from the Health Foundation (Registered Charity Number: 286967) and the National Institute for Health Research Health Services and Delivery Research (NIHR HS&DR) programme (project number 10/1007/26). Visit the HS&DR website for more information. The views and opinions expressed therein are those of the authors and do not necessarily reflect those of the HS&DR programme, NIHR, NHS or the Department of Health.

Chapter 16

When disaster strikes

Sustained resilience performance in an acute clinical setting

Lev Zhuravsky

Introduction

The Canterbury region hosts New Zealand's second largest city, Christchurch, located on the edge of the Canterbury plains, bordered by the Port Hills and the Pacific Ocean (Christchurch City Council, 2013). In 2010, the city had a population of approximately 400,000, and experienced a high level of national and international tourism. While there are seven hospitals within the city, only one of these is a tertiary-level, acute-admitting facility with emergency department services. The other hospitals specialise in rehabilitation, mental health, non-acute medical and surgical services (Richardson and Ardagh, 2013).

The magnitude 6.3 Christchurch earthquake struck the city at 12.51 pm on Tuesday, 22 February 2011. The earthquake caused 185 fatalities, approximately 8,600 injuries and widespread damage to the built environment. The Christchurch earthquake badly damaged over 6,000 residential properties, forced thousands to leave their homes and communities, and disrupted the city's main lifelines including roads, water and wastewater networks, and electric distribution systems (Giovinazzi et al., 2011). This event compounded the effects of the magnitude 7.1 Darfield earthquake, which occurred on 4 September 2010 close to the rural town of Darfield. The earlier earthquake caused widespread property and infrastructure damage, but no direct fatalities.

The 22 February 2011 earthquake heavily impacted the Canterbury region's health care system. Christchurch Hospital sustained damage following the earthquake, which severely strained the hospital's ability to function at regular capacity. Due to water damage from leaking roof tanks, the top two floors of the 'Riverside' hospital building (including three adult medical wards) were evacuated immediately after the earthquake and relocated to an older person's health hospital. Relocating these wards was a major challenge for the entire general medical service.

Despite the damage, much of the hospital continued to function as normal. The acute medical wards on the 4th and 5th floor of the Riverside building were the only closures due to quake damage. Of the adult medical beds across the hospital, 106 were unusable. This amounted to a 16 per cent loss in total inpatients capacity (MacIntosh et al., 2012).

Rationale

This case study aims to demonstrate the practical application and positive contribution of the Resilient Health Care (RHC) approach (Hollnagel, 2012) to sustained nursing performance in one of the relocated wards during the initial 2 years following the earthquake. The team complemented the RHC model with other components based on their learning and experience. These included: shared leadership within the team and between charge nurse managers of all three relocated acute medical wards and interactive communication with medical and nursing teams across these three wards. Shared leadership in this particular case is defined as a system of management/leadership that involves all staff in the decision-making processes. Although not a novel invention, the concept of shared leadership has only recently been applied in the context of acute health care (St. Pierre et al., 2011).

This combined framework is similar to one described in the study conducted by van der Beek and Schraagen (2015) where shared leadership was viewed and described as a driver for all four essential capabilities of resilience together with comprehensive communication.

As has been mentioned before, shared leadership does not constitute a part of the Resilient Health Care approach. Instead, it has been shown to positively impact team resilience in acute clinical settings in the context of evolving crisis. This concept will be described in a subcase involving shared leadership in the Christchurch Hospital Intensive Care Unit following the Christchurch earthquake using outcomes of qualitative research.

Data sources and data collection

The current case study was developed from the experiences and reflections of the author, who was a charge nurse manager in an adult medical ward during the Christchurch earthquake. He led a nursing team for 2 years after evacuation and relocation to new premises at an older person's health hospital. Autoethnography is introduced as the proposed methical scaffolding for this particular case study.

Autoethnographic methodology falls into a continuum of ethnographic research approaches that serve to describe, interpret, analyse and illuminate phenomena in fieldwork studies, geocultural landscapes and bounded case study frameworks. But this methodology departs from, and challenges, traditional ethnographic research because (a) the researcher's intrapersonal perspective is the focus of the research, (b) it evolves through a reflexive process that occurs through creative writing and other arts-based techniques, and (c) it maintains that transformation occurs through active engagement with the material, rather than from generating external discussions and generalisations in a broader context. The researcher is immersed in the reflexive process as well as positioned as a scholar-practitioner, and the required agility for such effort may also facilitate the ability to straddle different cultural realities simultaneously (Woodward, 2015).

The data collection process comprised two main components. The first component consisted of creating descriptive narrative text. The second component included the gathering of reflective notes.

Background

Department of general and internal medicine of Christchurch Hospital has close links with other internal medical specialties including health care of the elderly. In 2010, the department handled more than 9,000 admissions and comprised eight internal medical wards to care for this large number of patients. This included one respiratory and two cardiology/nephrology wards with the general wards taking subspecialties of gastroenterology, rheumatology, infectious disease and stroke.

In 2010 the department had a bed base of 241. General medicine is the 'default' admitting service for the hospital. This means that if there is any doubt about which service a patient should be admitted to then the patient is accepted by general medicine.

On 22 February 2011, three acute medical wards were evacuated and permanently closed in the Riverside building of Christchurch Hospital as the result of the post- earthquake decision to limit inpatient occupancy of floors without horizontal evacuation options. These wards were moved to temporary accommodation at the older person's health hospital – The Princess Margaret Hospital (TPMH). Two medical wards remained on Christchurch's public hospital (CPH) campus. This split of the acute services has had a large impact on models of care and delivery of service within general medicine.

Ward 29's team of nursing and support staff, which is described in this case study, were the first to relocate. At the time, this was a 27 bedded acute general medical ward. The team included Charge Nurse Manager (CNM), 31 registered nurses, 10 health care assistants and 3 clerical staff. Relocation took place within a week of the earthquake and involved complex planning and intricate execution, compounded by ongoing seismic activity. The team had to operate under tight time constraints and were expected to become fully operational with new admissions within 3 days of relocation. The ward space the team moved into lacked proper infrastructure but within a short period of time IT systems were operating, acute admitting medical teams were formed and modified policies and protocols were agreed on that reflected required operational changes. For 2 months the team remained the only acute general medical team at TPMH premises. Due to a lack of acute medical beds at Christchurch hospital, two remaining medical wards were relocated to TPMH later on to form a small cluster of acute medical services based at TPMH.

In these wards, for more than 2 years medical and nursing teams continued to provide acute medical care to a diverse population under difficult conditions where they coped with complex clinical and operational challenges and continuous seismic activity. These challenges required all teams to develop and maintain a high level of resilience. Performance of Ward 29 team fits the Resilient Health Care model through integration of the following components of the model into Everyday Clinical Work (ECW):

- Four core capabilities: anticipating, monitoring, responding, and learning.
- Workarounds as an integral part of a capability to respond.
- Realignment of Work-as-Imagined (WAI) with Work-as-Done (WAD).

The RHC approach has the potential of enhancing overall resilience at team level and the ability to contribute to development of new creative solutions in the context of evolving crisis. Also, RCH has a strong potential to improve and support individual resilience and personal capacity to cope and adapt.

As mentioned before, the RHC approach was complemented by adopting a model of shared leadership within the nursing team and by establishing good communication with medical and nursing teams in two other medical wards and acute teams at Christchurch Hospital.

Analysis and interpretation

Collected data have been analysed using deductive qualitative analysis where existing research and theory paves a way to a study. Sources of the theory to be tested come from previous research and personal and professional experience.

Resilience can be defined as the ability of the health care system (a clinic, a ward, a hospital, a county) to adjust its functioning prior to, during, or following events (changes, disturbances and opportunities), and thereby sustain required operations under both expected and unexpected conditions (Hollnagel, Wears and Braithwaite, 2015).

Hollnagel (2009b) proposed a list of four capabilities that enable and promote resilient performance. These include the:

- Capability to respond. A system must respond in an appropriate manner in real time.

- Capability to monitor. An ongoing search to identify threats and opportunities.
- Capability to anticipate. A forward-looking approach, anticipating what could happen and preparing responses.
- Capability to learn. A multi-level learning process evaluating what went wrong and what went right.

Capability to respond

Initially the Christchurch earthquake required a quick response from all the team's members. All beds on Ward 29 were occupied and some of our patients required complex acute care. Because of ongoing aftershocks, staff could not stand unsupported, and although there was no structural damage the physical environment of the ward was significantly compromised. While providing care to patients, the staff on the ward had to manage their personal stress, fear of the unknown and concerns about their families and relatives. It should be noted that the team had no prior training in responding to a crisis event of this type and scale. Initially, part of response was based on self-organising behaviour and leadership. During that day the response focussed on three main things: first, patient safety and care; second, safety of the environment; and, finally, preparing for possible evacuation.

Leadership played an important part in the response. New leadership structures quickly were established, including shared leadership where informal leaders and members of medical teams and other senior members of nursing teams cooperated and coordinated in a shared decision-making process.

The situation became more challenging after Ward 29 was relocated. The variety of complex problems faced while establishing and running the fully operational and viable acute medical word required prompt solutions and 'out of the box' thinking. To sustain resilience performance over time the team were up-skilled and trained. There were numerous simulation and debrief exercises and new operational protocols and procedures were established. Simulation and debriefings were mainly focussed on what the team did well and on what worked. To sustain and support uninterrupted acute admissions and patients flow, a new model of care has been developed and implemented within a very short time frame. At the centre of the new model is a process of direct acute admissions into three relocated wards. Instead of accepted clinical pathway where patients from community used to come first to the Emergency Department and then got admitted into inpatients wards, the new process allows patients to be admitted into three medical wards at TPMH directly from the community, bypassing the Emergency Department. The new process required ongoing training and support for all members of the clinical teams and over time they developed a high degree of adaptability and capability to respond.

The concept of 'workaround' also contributed to the team's capability to respond well. Workaround is defined as circumventing or temporarily fixing work flow used to achieve goals readily by avoiding perceived hindrances (Debono et al., 2013). Workaround facilitates practice to continue against adversity, enabling people to accommodate challenges and maintain effective delivery of care (Tucker, 2009).

Ward 29's approach was to make workarounds part of their operational strategy. Initially the team recognised what was important and agreed on open and transparent ways of reporting and learning from this. Gradually they integrated workarounds into their daily operations.

Innovative and flexible solutions increased the adaptive capacity of the team, and helped to realign a gap between Work-as-Imagined and Work-as-Done. This eventually contributed to Ward 29's capability to respond and helped sustain the team's resilience performance over time.

Capability to monitor

'Capability to monitor' was one of the key contributing factors in creating the robust and sustained response of the team. After relocating, it was important to monitor a number of threats and identify suitable opportunities. First of all, the Ward 29 team had to monitor indicators related to clinical performance to ensure that they were providing safe and timely care. Situations they faced and had to cope with were unusual and required continuous input. Patients who were acutely admitted to relocated wards had to be constantly monitored and assessed to identify any early signs of deterioration. TPMH lacked an Emergency Department and advanced clinical support after hours. Therefore, early detection of any signs of clinical deterioration and prompt transfer of patients to Christchurch Hospital were a key element in ensuring continuous and comprehensive safe clinical care.

At the same time, it was vital to monitor how staff were coping with the pressure and ongoing challenges.

Proactive monitoring of signs of stress, fatigue and anxiety early helped to create an open and supportive environment, which in turn contributed to the overall team's resilience. Staff had a wide range of options with regards to where and how to seek help and support. Also, given the evolving nature of the event and ongoing aftershocks, the team had to continuously monitor their physical environment to identify and report any potential damage.

The split of acute services across two sites posed a significant risk to patients' safety and continuity of care. To mitigate this risk and to ensure comprehensive handover between clinical teams it has been decided to utilise advanced technical capabilities of the telemedicine and to run daily handover sessions for medical teams across two hospital sites.

Capability to anticipate

As soon as the team established robust mechanisms to monitor the physical environment, safety of patient care and well-being of staff, they started developing their capability to anticipate. The purpose of looking for what may potentially happen was to identify possible future events and changes in the environment that could have affected the team's ability to function. They had regular staff meetings and debrief sessions in smaller groups. These discussions contributed to their vision of what potentially might happen and the team's risk awareness improved.

Capability to learn

The capability to respond depends on the capability to monitor, in the sense that the timing and precision of responses can be improved by effective monitoring. However, the capability to respond and the capability to monitor both depend on the capability to learn (Hollnagel, 2015a).

Developing a robust and comprehensive learning structure was one of the main objectives of the team. Given the complexity of circumstances and wide range of unique challenges, conventional learning tools, known as single-loop learning, could not provide the required outcomes. Single-loop learning focusses on establishing rigid strategies, policies and procedures and then prompting people to spend their time detecting and correcting deviations from the rules.

Therefore, the team had to adopt a double-loop approach to learning where they were encouraged to develop creative and critical thinking. Research confirms that double-loop learning is critical to the success of teams, especially during times of rapid change.

To achieve the desired outcomes, the team needed enough opportunities to learn. Staff were encouraged to share their experiences in different ways. Teaching sessions were run for small groups that encouraged personal reflections and specific attention to debriefing processes. Everything the team did turned into learning processes with many opportunities to translate what they had learnt into practice.

Realignment of Work-as-Imagined (WAI) with Work-as-Done (WAD)

In resilient engineering terms, the blunt end of the system is usually called Work-As-Imagined (WAI) and the sharp end is called Work-As-Done (WAD). In a complex adaptive system like a health care team or organisation, WAD on a frontline of patient care is always different from WAI by those who manage and lead organisational clusters and units (Hollnagel, Braithwaite and Wears, 2013c).

Research has found that practitioners adapt their behaviour based on their experience and expertise. The resulting behaviours are an example of how Work-As-Done (WAD) by practitioners is necessarily different from Work-As-Imagined (WAI) by systems designers and managers (Hollnagel et al., 2013c). If WAI and WAD do not correspond to each other, the solution is to reduce the misalignment by realigning the two. The idea of realignment acknowledges that neither WAI nor WAD is an absolute reference, and that the gap between them should be reduced to facilitate organisational resilience.

A process of ongoing realignment between WAI and WAD at team level played a very important role in ensuring resilience performance. Establishing clear communication across a team, transparency, trust and open engagement assisted in turning this process into a learning opportunity for a whole team. Collaboration, innovation and engagement between team leaders and other team members contributed to an overall team's resilience.

Shared leadership – Acute medical ward

At the time of the earthquake and following relocation, the Ward 29 team quickly adopted a shared leadership approach to decision making and management. At a very early stage they had a strong group of emergent leaders. An emergent leader is someone who is not designated as a leader, but emerges as an informal leader of the group by exerting influence on group processes and group goal achievement.

As an example, in 2005 Hurricane Katrina created a situation where the capacity of formal leaders to handle the situation was overwhelmed. It created a clear void in leadership structure. To fill this void, individuals stepped up to assist others and informal leadership capabilities came to the fore.

In contrast to Hurricane Katrina, Ward 29 staff were motivated to step up and stretch themselves during the crisis in Christchurch by a desire to support their team at many levels to keep things going at a very challenging time. Shared decision making and engaging formal and informal leadership groups within the team during the response contributed to the team's overall resilience. It also assisted in realigning the gap between WAI and WAD. True shared leadership was based on mutual trust and open interaction at multiple levels within the team. One of our main learnings from this process of engagement with informal leaders was that:

- Successful performance in a crisis situation provides opportunity for informal leaders to step up. Support of informal leaders begins with understanding how they approach different tasks during a crisis situation.

The team had the ability and the right tools to assist informal leaders in creating a positive and motivating environment. Informal leaders in the team needed to know that their positive attitude and creative contributions were of great value to the team and organisation. Recognising efforts and new opportunities can be one of the most highly valued forms of reward. Informal leaders were an essential component in responding to the challenges that Ward 29 had to cope with and their contribution to successful performance of the team cannot be underestimated. Collaboration, trust and a high level of management support contributed to the ability of the staff to exercise informal leadership.

Shared leadership – Intensive Care Unit, Christchurch Hospital

Teams within health care organisations often need to respond to sudden, unanticipated demands and then return to normal operating conditions as quickly as possible and with a minimum of performance loss. Preliminary findings from research on resilience following the Canterbury earthquakes indicate that adaptive resilience, that is, active responding in the face of crises, requires shared leadership.

Internationally, a shared leadership model could be understood as a system of management/leadership that empowers all staff in the decision-making processes. Published research (McCallin 2003; Ensley, Hmieleski and Pearce, 2006) has described shared leadership as an interdisciplinary team leadership model, a form of distributed leadership originating within teams operating in a high stakes clinical environment.

The nature of a crisis makes it difficult to study the role of shared leadership in a comprehensive and systematic manner. By definition, crises are rare and complex events.

Research cannot possibly predict when and where a crisis will occur and which people will become leaders. A challenge of crisis leadership is to enhance both individual and collective leadership (Rego and Garau, 2007). In a crisis situation individual leadership is imperative but often insufficient.

Shared leadership approaches have demonstrated better outcomes than individual leadership in a variety of contexts, and are considered to be particularly effective in situations of complexity (Knox, 2013). Collective leadership means everyone taking responsibility for the success of the organisation as a whole – not just for their own jobs or work area. This contrasts with traditional approaches to leadership, which have focussed on developing individual capability while neglecting the need for developing collective capability or embedding the development of leaders within the context of the organisation they are working in (West et al., 2014).

The Intensive Care Unit (ICU) within Christchurch Hospital is a specially staffed and equipped department treating acutely ill or injured patients with life threatening disorders. The ICU employs seven consultants and more than 100 nurses supported by a Unit Manager and senior nursing staff. The total physical capacity of the ICU is 18 acute beds but only 12–15 are usually staffed. When the earthquake struck, 15 of the 18 available ICU beds were staffed to take patients requiring acute life support. Fourteen beds were in use at the time of the earthquake. The first patient arrived in ICU within an hour of the earthquake. It immediately became clear that most of the existing patients needed to be transferred out of ICU to create more critical space for earthquake victims.

The author undertook a qualitative study that determined if shared leadership was possible and warranted during a crisis caused by a natural disaster. The study assessed the nature of specialist and nursing leadership in the Intensive Care Unit of Christchurch Hospital within the first 72 hours after the earthquake (Zhuravsky, 2015).

The results from the study showed that during a crisis the team in the Intensive Care Unit adopted a shared leadership approach, which comprised two main elements. The first element was sharing of leadership within the formal leadership group, which includes both medical and nursing sub groups. The second element was sharing of leadership between formal and informal leaders in the unit. Shared leadership as experienced and described by the participants in the study was close to the concept developed by Mielonen (2011) where shared leadership is defined as not just a new practical arrangement but a process of working together. The process required sharing power, authority, knowledge and responsibility. If people are to work together genuinely, they need to engage fully in the realities of problem solving and decision making of leadership tasks and be empowered to act with a certain degree of authority. Most of the participants in the study indicated that sharing leadership tasks assisted in creating a coordinated unit wide response and provided a good framework for decision making and overall management of the crisis. It reduced task overload and increased team performance.

Sharing of leadership in a crisis can create a synergy of expertise where formal and informal leaders utilise their individual strengths and bring diversity into the process of crisis management. As reported by some participants, reduced stress levels for the leadership group makes this approach more attractive because a more robust, shared leadership approach does not unduly burden any single leader. The results of this study provide some evidence for adopting a shared leadership model in crisis. It is in line with other research where shared leadership was found appropriate in situations with high task load induced by a non-routine event (Kunzle et al., 2010).

Conclusion

The purpose of this chapter was to discuss the contribution of a Resilient Health Care model and its concepts of responding, monitoring, learning and anticipating to the sustained resilient performance of the nursing team at one of the medical wards of Christchurch Hospital over the period of 3 years following Christchurch earthquake. In particular, this case study focussed on capability to learn as a main contributing factor to the resilience continuum, contribution of workarounds to a team's resilience, importance of realigning the gap between Work-as-Imagined and Work-as-Done and to exploring the link between shared leadership and an overall team's resilience.

The Ward 29 team, like many other teams in the Canterbury District Health Board, demonstrated a remarkable ability to go beyond their original level of functioning and to grow and thrive despite repeated and prolonged exposure to stressful experiences.

Chapter 17

Making it happen – from research to practice

Erik Hollnagel and Jeffrey Braithwaite

Introduction

The noble purpose of research is to consolidate the knowledge – the collection of theories, models, hypotheses, experiences and assumptions – that constitutes the basis for how we interact with the world around us. This knowledge determines what we pay attention to, what we look for, how we make sense of what we see and what we do to make sure that our actions will have the intended outcomes. This purpose applies for research in the established sciences as well as for research in fields that may be on their way to become sciences, such as resilient health care (RHC).

Chapter 3 described how the emphasis on RHC at the present time is more on gathering evidence and formulating hypotheses than on testing hypotheses or confirming models. This was further illustrated by the 13 research chapters in this book. The chapters all address issues of a practical rather than a speculative nature, hence they represent applied rather than pure or basic research. The motivation comes both from the experiences with everyday clinical work and from the inevitable frustrations in changing and improving it that together express a need of knowledge that may lead to practical solutions in both the short term and the long term.

A central question is how the research findings described here can be used to change how health care systems work, whether it be on the micro, the meso or the macro level. The chapters did not directly address this issue, partly because there is a traditional distinction between research itself and the use of the results. One reason for this is that the two require different types of rather specialised skills. Another is that they fall within the remit of different parts of an organisation or even take place in completely different organisations. Research groups in universities, for example, are not always closely associated with provider organisations. Research and the use of the results also differ with regard to how well they are structured and the time they take. Research usually has a well-defined plan or study protocol with a clear beginning and end while practical changes based on the research findings are more open ended. It may take years or even decades before results can be seen – and they may not necessarily be what the proponents had intended. Although this chapter cannot make amends for that, it can at least identify some of the most important aspects of adopting research findings.

Planned changes and emergent changes

The problems of how to change practice, i.e., how to change the ways in which an organisation and the people in it work, are of course not new but have been a challenge to leaders and administrators for thousands of years. Up until the industrial revolution in the second half of the eighteenth century, the need to change came mostly from revolutionary technologies (e.g., movable type) or political upheavals and was therefore few and far between. After the industrial revolution and the appearance of industrial societies, the growing dependence on technology in production, transportation, services, etc., meant that changes were needed more often, usually due to improvements of existing technologies but at times also because of the appearance of new technologies. During the last 60 years or so, the rate of change has continued to grow leading to conditions where hitherto applicable solutions no longer suffice.

The traditional approach to planned change is predicated on a paradigm that sees change as a transitional process between fixed states. The starting point is a system in a steady state that works, but where there is a need of improvements either in the form of new technology or new ways of organising work. This leads to a period of transition during which the improvements are introduced and brought to bear on the established practice. The transition period lasts until a new steady state has been achieved, although the time needed for that rarely is correctly estimated – if known at all. When the change has been completed, the system begins to function in the new steady state until another change is required. The changes can either be gradual, such as improved technologies that make it possible to do something a little better, faster or cheaper, or radical, for instance by replacing humans with robots or by adopting totally different ways of working such as replacing invasive surgery with laparoscopic techniques.

When the industrial society emerged, the drivers of changes were mainly the expected benefits of new technology. To this was soon added the need to improve efficiency, quality and competitive performance. In consequence of that, changes came to be driven less by new technology than by new ways of organising work and business, i.e., by a need to alter the structure or type of organisations. The approach, however, was the same as illustrated by Lewin's classic three-stage change model of 'unfreezing, change, and refreezing' (Lewin, 1952). The purpose of the 'unfreezing' stage is to create an awareness that the *status quo* in some way limits the organisation and that a change therefore is needed. The 'changing' stage is the transition from one way of working to another, and may involve the assimilation of new technologies. It is during the changing stage that people adopt new behaviours, adjust to new conditions and hopefully get used to new ways of thinking. The third and final stage is the 'refreezing' – or just 'freezing' – where the changes gradually become institutionalised and therefore are accepted as the new 'normal'. One purpose of the 'refreezing' is to prevent people from reverting to old habits once the transition phase is over since that would nullify the effects of the change.

The planned changes approach works fine as long as two important assumptions are fulfilled. There must first of all be a good understanding and description of how the system works, which traditionally means that it is possible to account for the system using linear (cause-effect) thinking. Indeed, the very idea of a transition from one steady state to another represents a linear, causal paradigm. The second assumption is that the system can be in a steady state, both before the change begins and after it has been completed. Although they rarely have been stated explicitly, the two assumptions were quite reasonably taken for granted throughout the nineteenth and most of the twentieth century. But around the 1970s things started to change. Organisations developed in ways that could not always be foreseen due to increasing vertical and horizontal integration and because the conditions under which they operated went from being reasonably stable and predictable to become more increasingly unstable and unpredictable. Although a planned change eventually could move an organisation from one steady state to another, it never happened as imagined. One the one hand the intended consequences would often not be obtained (something that had been recognised already by Merton, 1936), and on the other there would be unexpected consequences, positive as well as negative.

A possible third assumption, recognised by Lewin's description of 'unfreezing', is that the people who will be affected by the mooted change are receptive to it and willing to embrace it. During the second half of the twentieth century it was gradually realised that frontline practitioners were not passive recipients of change, but had what philosophers and sociologists refer to as agency (Sewell, 1992). That is, people can exert some degree of control over the social relations in which they are enmeshed, which in turn implies the ability to transform those social relations to some degree. People have varying degrees of independence of action and thought, reflexivity, voice, varying practices and levels of autonomy. This altogether led to the idea of emergent changes, where changing an organisation was seen less as a prescriptive and more as an analytical undertaking. The concept of emergent changes decreases the importance of plans and projections and increases the importance of understanding the complexity of the operating environment and the ability to make adjustments during the transition (Braithwaite et al., 2017).

Implementation

In the health care systems and organisations of today, the need to change typically arises before an already ongoing change has been completed, either within the same part of the organisation or in a different but related part. The need to change can be due to technological innovations, new forms of treatment, regulatory demands, management initiatives, economic conditions, etc. A change therefore rarely starts from a stable state, nor does it take place in a stable environment. The situation is obviously made more difficult if yet another change is introduced before the previous one has been completed, and so on. (This often happens because the previous change turned out to have unexpected and unwanted side effects.) Yet the tools of change management were developed for situations where there was a well-defined change to make, and clear, specific expectations to the outcome or results. These conditions are not met in current health care systems, nor are they met by the issues that are at the forefront of resilient health care. In reality, changes envisaged in one part of the system, or for a specific purpose, may directly or indirectly interact with other changes during various stages of the implementation.

A proposed solution to these problems – which of course are not peculiar to health care – has been a more systematic way to translate research findings into practical called 'implementation science'. Despite the good intentions, the results have so far been rather disappointing. 'Implementation science ... is contested and complex, with unpredictable use of results from routine clinical practice and different levels of continuing assessment of implementable interventions' (Rapport et al., 2017). The reason for that can be found by looking at the assumptions made by implementation science. A systematic review of the relevant literature argued that an intervention goes through the five phases shown in Figure 17.1 (Braithwaite, Marks and Taylor, 2014).

The five phases are easily recognisable as a variation of the planned changes paradigm and as an extension of Lewin's three-step model. The five phases are assumed to happen one after the other, hence represent a linear development. They also assume that the implementation begins and ends with the system being in a steady state and that changes are made sequentially rather than simultaneously. Since it is unlikely that any of these conditions will be true in present-day health care systems, it is hardly surprising that simply following step-by-step guides for implementation runs into problems. This model, and others like it, is therefore more an idealisation than a representation of the complexity of health care organisation (Braithwaite et al., 2017).

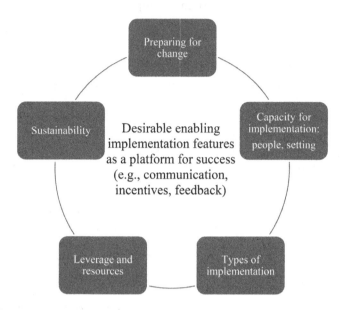

Figure 17.1 Phases of implementation.
Source: Braithwaite, Marks and Taylor (2014).

The nature of changes

Many research findings will inevitably point to changes that are 'intangible' in the sense that they refer to issues related to the perception of work, social relations, attitudes and even the politics and culture, rather than specific work components, tasks or 'gadgets'. The changes can be about doing things in a different way, but are more likely to be about looking at what we do in a different way and through that change how things are done. Resilient health care is less about different or new practices and more about a different or new way of looking at practice. Examples taken from the chapters in this book are:

- A need to better understand when workarounds contribute to resilience and when they do not.
- A way to overcome knowledge gaps.
- Becoming able to disseminate resilience strategies more widely.
- Supporting an organisation's ability to diagnose itself and understand factors that contribute to both success and failure.
- Supporting an iterative and emergent process of engagement and dialogue with a practice community.
- Reducing unnecessary adjustments at the sharp end.

Even though such changes may be seen as 'intangible' they are nevertheless essential in order to strengthen an organisation's potentials to perform in a resilient manner. Setting wholesale solutions such as improving patient safety culture aside, researchers, managers and frontline staff need to come together to develop practical ways to bring resilient health care into practice. Recognising the fact that we live and work in a world that is unstable and unpredictable, there will clearly be no simple recipes for how to implement changes. The starting point should instead be the acknowledgement of the existing conditions, as described above, and in particular that changes must take place in turbulent and incompletely controlled conditions. The overall guidance is therefore to be thorough rather than to be efficient (Hollnagel, 2009a), by paying attention to the following points.

Before thinking about how to make a change, it is wise to invest time to find out how the system actually functions and how everyday situations and uncertainties are successfully managed. In other words, make sure that the changes are based on an evidential and thorough understanding of WAD rather than a convenient description of WAI.

Try to be as detailed as possible in describing the intended outcomes, what you expect to be different after the change has been made. Will there be directly observable changes in how work is done, or even measurable changes in what is produced? Or will the changes rather be to attitudes, the way politics are practised, how cultural features manifest themselves, or to social relations? Which aspects of work will be different and why? Having as many details as possible will make it easier to determine whether the change has been successful. People are often overly optimistic about what they will find and how much they will accomplish and may therefore be biased in their judgement of whether they have been successful.

Think about how you will keep track of how the change takes place. In other words, how progress can be monitored and/or measured. It is not sufficient to rely on outcome measures such as end stage performance indicators, since these only cover part of the change and furthermore do not offer the possibility of intervening and adjusting the change while it happens. Keep in mind that changes in today's systems are emergent rather than planned. Since there are few if any measures of organisational processes, one possibility is to make use of proxy measures instead. The RAG or Resilience Analysis Grid, which has been used in several of the studies, may be a candidate for that.

Finally, try to be realistic about how long it will take for a change to happen. This is often overlooked, even in the case of changes that involve technology. It is, of course, possible to issue a new guideline or recommendation from one day to the next, but it is difficult to estimate how much time it takes before it becomes accepted practice. Speaking in completely general terms, it is more likely to take months rather than weeks, and possibly even years rather than months. It is necessary to have a reasonably realistic idea about the duration of a change both to know how to monitor it and to know when to look for the results. Do not discount the possibility that the results may first be seen after the tenure of the current change agents have come to an end.

The bottom line is that the understanding of how to move from research to practice in clinical settings is still in its infancy. Bringing research results into practice is therefore perhaps less an issue of managing change than an issue of changing managing.

References

Aase, K. (2010). Pasientsikkerhet – Hendelser, begreper og omfang. In: K. Aase (Ed.), *Pasientsikkerhet-teori og praksis i helsevesenet* (Patient safety in healthcare – theory and practical). Oslo, Norway: Universitetsforlaget.

Aase, K. and Wiig, S. (2010). Skape og opprettholde et lærende helsevesen? In: K. Aase (Ed.), *Pasientsikkerhet-teori og praksis i helsevesenet* (Patient safety in healthcare – theory and practical). Oslo, Norway: Universitetsforlaget.

Amalberti, R. et al. (2011). Adverse events in medicine: Easy to count, complicated to understand, and complex to prevent. *Journal of Biomedical Informatics, 44*(3), 390–394.

Anderson, J. E. et al. (2014). Operationalising resilience in healthcare: Theory, methods and data. *The Resilient Health Care Net Summer Meeting*, 12–14 August, Middelfart, Denmark.

Anderson, J. E. et al. (2016). Implementing resilience engineering for healthcare quality improvement using the CARE model: A feasibility study protocol. *Pilot and Feasibility Studies, 2*(61).

Anderson, J. E., Ross, A. and Jaye, P. (2013). *Resilience engineering in healthcare: Moving from epistemology to theory and practice.* In: 5th Resilience Engineering Association Symposium on Resilience Engineering. Managing Trade Offs. June 25–27, Soesterberg, The Netherlands.

Anderson, J. E., Ross, A. J. and Jaye, P. (2016). Modelling resilience and researching the gap between Work-As-Imagined and Work-As-Done. In: J. Braithwaite, R. L. Wears, R. L. and E. Hollnagel (Eds), *Resilient health care, vol. 3: Reconciling work-as-imagined with work-as-done*. Farnham, UK: Ashgate.

Anderson, L. (2011). Use the world café concept to create an interactive learning environment. *Educ Prim Care, 22*(5), 337–338.

Apkon, M. et al. (2004). Design of a safer approach to intravenous drug infusions: failure mode effects analysis. *Qual Saf Health Care, 13*, 265–271.

Ashby, W. R. (1956). *An introduction to cybernetics.* New York, NY: Wiley.

Australian Council for Safety and Quality in Healthcare Medication Safety Task-force. (2003). Intravenous potassium chloride can be fatal if given inappropriately. Medication Alert! *Medication Alert 1.*

Azad, B. and King, N. (2012). Institutionalized computer workaround practices in a Mediterranean country: an examination of two organizations. *Eur J Inf Syst, 21*, 358.

Ball, D. R. and Frerk, C. (2015). A new view of safety: Safety 2. *British Journal of Anaesthesia, 115*(5), 645–647.

Bates, D. et al. (2005). American College of Endocrinology and American Association of Clinical Endocrinologists position statement on patient safety and medical system errors in diabetes and endocrinology. *Endocr Pract, 11*, 197–202.

Becker, A. B. et al. (1995). Effects of jet engine noise and performance feedback on perceived workload in a monitoring task. *Int J Aviat Psychol, 5*, 49–62.

Berg, M. (1994). Modeling medical work: on some problems of expert systems in medicine. *SIGBIO Newsletter, 14*, 2–6.

Bodenheimer, T. and Sinsky, C. (2014). From triple to quadruple aim: Care of the patient requires care of the provider. *Ann Fam Med, 12*, 573–576.

Bourdieu, P. (1977). *Outline of a theory of practice.* Cambridge: Cambridge University Press.

Bourdieu, P. (1984). *Distinction: A social critique of the judgement of taste.* Cambridge, MA: Harvard University Press.

Bourdieu, P. (1990). *The logic of practice.* Stanford, CA: Stanford University Press.

Braithwaite, J. et al. (2013). Health care as a complex adaptive system. In: E. Hollnagel, J. Braithwaite and R. L. Wears (Eds), *Resilient health care.* Farnham: UK, Ashgate, pp. 57–76.

Braithwaite, J. et al. (2017). *Complexity science in healthcare: A White Paper.* Sydney, Australia: Australian Institute of Health Innovation, Macquarie University.

Braithwaite, J., Marks, D. and Taylor, N. (2014). Harnessing implementation science to improve care quality and patient safety: A systematic review of targeted literature. *International Journal for Quality in Health Care, 26*(3), 321–329.

Braithwaite, J., Runciman, W. B. and Merry, A. F. (2009). Towards safer, better healthcare: Harnessing the natural properties of complex sociotechnical systems. *Quality and Safety in Health Care, 18*(1), 37–41.

Braithwaite, J., Wears, R. L. and Hollnagel, E. (2015). Resilient health care: Turning patient safety on its head. *Int J Qual Health Care, 27*(5), 418–420.

Braithwaite, J., Wears, R. L. and Hollnagel, E. (Eds) (2017). *Resilient health care, vol. 3: Reconciling Work-as-Imagined with Work-as-Done.* Farnham, UK: Ashgate.

Braun, V. and Clarke, V. (2006). Using thematic analysis in psychology. *Qualitative Research in Psychology, 3*, 77–101.

Brown, J. and Isaacs, D. (2005). *The World Café: Shaping our futures through conversations that matter.* San Francisco, CA: Berrett-Koehler Publishers.

Burnett, S. et al. (2013). Prospects for comparing European hospitals in terms of quality and safety: lessons from a comparative study in five countries. *International Journal of Quality in Health Care, 25*(1), 1–7.

Carpenter, C. (2010). Phenomenology and rehabilitation research. In: P. Liamputtong (Ed.), *Research methods in health foundations for evidence based practice.* Oxford: Oxford University Press, pp. 123–140.

Carter, S. M. and Little, M. (2007). Justifying knowledge, justifying method, taking action: Epistemologies, methodologies, and methods in qualitative research. *Qualitative Health Research, 17*(10), 1316–1328.

Catchpole, K. R. et al. (2007). Improving patient safety by identifying latent failures in successful operations. *Surgery, 142*, 102–110.

Catchpole, K. R. and Russ, S. (2015). The problem with checklists. *BMJ Qual Saf*, *24*, 545–549.

Chaffee, M. W. and McNeill, M. M. (2007). A model of nursing as a complex adaptive system. *Nursing Outlook*, *55*(5), 232–241.

Cheung, D. S. et al. (2010). Improving handoffs in the emergency department. *Ann Emerg Med*, *55*, 171–180.

Chiarini, A. (2013). Waste savings in patient transportation inside large hospitals using lean thinking tools and logistic solutions. *Leadership in Health Services*, *26* (4), 356–367.

Chopra, V. et al. (1992) Reported significant observations during anaesthesia: A prospective analysis over an 18-month period. *British Journal of Anaesthesia*, *68* (1), 13–17.

Christchurch City Council (2013). *Christchurch. The garden city*. Available at: www. christchurch.org.nz/(accessed 15 December 2015).

Clegg, C. (2000). Sociotechnical principles for system design. *Applied Ergonomics*, *31*, 463–477.

Clement, S. et al. (2004). Management of diabetes and hyperglycemia in hospitals. *Diabetes Care*, *27*, 553–591.

Committee on the Future of Emergency Care in the United States Health System (2007). *Hospital-based emergency care: at the breaking point*. Future of Emergency Care Series. National Academy Press, Washington, DC.

Conklin, T. (2012). *Pre-accident investigations: An introduction to organizational safety*. Farnham, UK; Burlington, VT: Ashgate.

Cook, R. I. (2013). Resilience, the Second Story, and progress on patient safety. In: Hollnagel, E., Braithwaite, J. and Wears, R. L. (Eds), *Resilient health care*. Farnham, UK: Ashgate.

Cook, R. I. and Ekstedt, M. (2016). Reflections on resilience: Repertoires and system features. In: J. Braithwaite, R. L. Wears and E. Hollnagel (Eds), *Resilient health care, vol 3. Reconciling work-as-imagined and work-as-done*. Burlington, VT: Ashgate.

Cook, R. I. and Rasmussen, J. (2005). 'Going solid': A model of system dynamics and consequences for patient safety. *Qual Saf Health Care*, *14*(2), 130–134.

Cook, R. I., Render, M. and Woods, D. D. (2000). Gaps in the continuity of care and progress on patient safety. *BMJ*, *320*, 791–794.

Cook, R. I., Woods, D. D. and Miller, C. (1998). *A tale of two stories: Contrasting views of patient safety*. Chicago, IL: National Patient Safety Foundation.

Cooperrider, D. L. and Srivastva, S. (1987). Appreciative Inquiry in organizational life. In: R. W. Woodman and W. A. Pasmore (Eds), *Research in organizational change and development, vol. 1*. Greenwich, CT: JAI Press, pp. 129–169.

Courtenay, M., Carey, N., James, J., Hills, M. and Roland, J. (2007). An evaluation of a specialist nurse prescriber on diabetes inpatient service delivery. *Practical Diabetes International*, *24*(2), 69–74.

Creswell, J. (2007). *Qualitative inquiry and research design: Choosing among five approaches*. Thousand Oaks, CA: Sage.

Creswell, J. W. and Miller, D. L. (2000). Determining validity in qualitative inquiry. *Theory into Practice*, *39*(3), 124–130.

Croskerry, P. (2009). *Patient safety in emergency medicine*. Philadelphia, PA: Wolters Kluwer Health/Lippincott Williams & Wilkins.

Cuvelier, L. and Falzon, P. (2008). Methodological issues in the quest of resilience factors. In: E. Hollnagel, F. Pieri and E. Rigaud (Eds), *3rd International Symposium on Resilience engineering*. 28–30 October 2008, France.

Davies, M. et al. (2001). Evaluation of a hospital diabetes specialist nursing service: a randomised controlled trial. *Diabetic Med, 18,* 301–307.

Dean, B. and Barber, N. (2001). Validity and reliability of observational methods for studying medication administration errors. *American Journal of Health Systems Pharmacy, 58,* 54–59.

Debono, D. and Braithwaite, J. (2015). Workarounds in nursing practice in acute care: A case of a health care arms race? In: R. L. Wears, E. Hollnagel and J. Braithwaite (Eds), *The resilience of everyday clinical work*. Farnham, UK: Ashgate.

Debono, S. et al. (2013). Nurses' workarounds in acute healthcare setting: A scoping review. *BMC Health Services Research, 13,* 175.

Dekker, S. W. A. et al. (2008). *Resilience engineering: New directions for measuring and maintaining safety in complex systems. Final report*. Lund School of Aviation.

Delaney, C., Daley, K. and Lajoie, D. (2006). Facilitating empowerment and stimulating scholarly dialogue using the world cafe model. *J Nurs Educ, 45* (1), 46.

Deming, W. E. (1994). *The new economics for industry, government, education*. New York: The MIT Press.

Deutsch, E. S. (2011). Simulation in otolaryngology: Smart dummies and more. *Otolaryngology – Head & Neck Surgery, 145*(6), 899–903.

Diabetes UK. (2009). *Position statement: Improving inpatient diabetic care – what care adults with diabetes should expect when in hospital*. London: Diabetes UK.

Diabetes UK. (2010). *Diabetes in the UK 2010: Key statistics on diabetes*. London: Diabetes UK.

Dietz, H. P. (2009). Our public health system: An accident waiting to happen?. *Medical Journal of Australia, 191*(6), 345–346.

Directorate of Health (2005). *... And it`s going to get better!* National Strategy for Quality Improvement in health and social services. Oslo.

Directorate of Health (2010). *A safe maternity care. Quality Requirements for maternity care*. Oslo.

Dixon-Woods, M. and Shojania, K. G. (2016). Ethnography as a methodological descriptor: the editors' reply. *BMJ Qual Saf, 25,* 555–556.

Durkheim, E. (2001). *The elementary forms of religious life*. Oxford: Oxford University Press.

Edmondson, A. C. (2012). *Teaming: How organizations learn, innovate, and compete in the knowledge economy*. Chichester, UK: John Wiley and Sons.

Eisenberg, E. M. (2008). The Social Construction of Healthcare Teams. In: C. P. Nemeth (Ed.), *Improving healthcare team communication: Building on lessons from aviation and aerospace*. Farnham, UK; Burlington, VT: Ashgate.

Elliot, M. and Liu, Y. (2010). The nine rights of medications administration: An overview. *British Journal of Nursing, 19*(5), 300–305.

Endsley, M. R. and Kaber, D. B. (1999). Level of automation effects on performance, situation awareness and workload in a dynamic control task. *Ergonomics, 42*(3), 462–492.

Ennen, C. S. and Satin, A. J. (2014). *Reducing adverse obstetrical outcomes through safety sciences.* Online at www.uptodate.com/contents/reducing-adverse-obstetri cal-outcomes-through-safety-sciences (accessed 31 January 2016).

Ensley, M., Hmieleski, K. and Pearce, C. (2006). The importance of vertical and shared leadership within new venture top management teams: Implications for the performance of start ups. *The Leadership Quarterly, 17*(3), 217–231.

Eppich, W. and Cheng, A. (2015). Promoting excellence and reflective learning in simulation (PEARLS): Development and rationale for a blended approach to health care simulation debriefing. *Simulation in Healthcare: The Journal of the Society for Medical Simulation, 10*(2), 106–115.

European Commission (2014). *Communication from the commission. On effective, accessible and resilient health systems.* Brussels.

Fairbanks, R. J. et al. (2014). Resilience and Resilience Engineering in Health Care. *The Joint Commission Journal on Quality and Patient Safety, 40*(8), 376–383.

Fetterman, D. (1998). *Ethnography: Step by step.* London: Sage.

Fiore, S. (2004). Process mapping and shared cognition: teamwork and the development of shared problem models. In: E. Salas and S. Fiore (Eds), *Team cognition: Understanding the factors that drive process and performance.* Washington DC: American Psychological Association, pp. 133–152.

Fisher, R. A. (1926). The arrangement of field experiments. *Journal of the Ministry of Agriculture of Great Britain, 33*, 503–513.

Flanagan, J. C. (1954). The critical incident technique. *Psychological Bulletin, 51*(4), 327–358.

Fletcher, G. et al. (2003). Anaesthetists' Non-Technical Skills (ANTS): Evaluation of a behavioural marker system. *Br J Anaesth, 90*(5), 580–588.

Flyvbjerg, B. (2011). Case study. In: N. Denzin and Y. Lincoln (Eds), *The Sage handbook of qualitative research.* Thousand Oaks, CA: Sage, pp. 301–316.

Frankl, V. E. (1985). *Man's search for meaning.* New York, NY: Simon and Schuster.

Furniss, D. et al. (2014). Unintentional non-adherence: can a spoon full of resilience help the medicine go down? *BMJ Quality & Safety, 23*(2), 95–98.

Furniss, D., Back, J. and Blandford, A. (2010). Resilience in emergency medical dispatch: Big R and little r. Presented at WISH (Workshop of Interactive Systems in Healthcare) in conjunction with CHI, 2010. http://discovery.ucl.ac.uk/594757/(accessed 23 December 2015).

Furniss, D., Back, J. and Blandford, A. (2011a). Unwritten Rules for Safety and Performance in an Oncology Day Care Unit: Testing the Resilience Markers Framework. *Proc. Fourth Resilience Engineering Symposium.*

Furniss, D., Back, J., Blandford, A., Hildebrandt, M., and Brogburg, H. (2011b). A resilience markers framework for small teams. *Reliability Engineering and System Safety, 96*(1), 2–10.

Gaba, D. M. (2007). The future vision of simulation in healthcare. *Simulation in Healthcare: The Journal of the Society for Medical Simulation, 2*(2), 126–135.

Geis, G. L. et al. (2011). Simulation to assess the safety of new healthcare teams and new facilities. *Simulation in Healthcare: The Journal of the Society for Medical Simulation, 6*(3), 125–133.

Gherardi, S. and Nicolini, D. (2003). To transfer is to transform: The circulation of safety knowledge. In: D. Nicolini, S. Gherardi and D. Yanow (Eds),

Knowing in organizations: A practice-based approach. Armonk, NY: M. E. Sharpe, pp. 204–224.

Giovinazzi, S. et al. (2011). Lifelines Performance and Management following the 22 February 2011 Christchurch Earthquake, New Zealand: Highlights of Resilience. *Bulletin of the New Zealand Society of Earthquake Engineering, 44*(4), 404–419.

Goodliffe, L. et al. (2014). Rate of healthcare worker–patient interaction and hand hygiene opportunities in an acute care setting. *Infect Control Hosp Epidemiol, 35,* 225–230.

Grant, R. W. et al. (2004).Trends in complexity of diabetes care in the United States from 1991 to 2000. *Arch Intern Med, 164,* 1134–1139.

Grant, S. and Humphries, M. (2006). Critical evaluation of appreciative inquiry. *Action Research, 4*(4), 401–418.

Greenfield, D. et al. (2011). Factors that shape the development of interprofessional improvement initiatives in health organisations. *BMJ Qual Saf, 20*(4), 332–337.

Greenhalgh, T., Russell, J. and Swinglehurst, D. (2005). Narrative methods in quality improvement research. *Quality and Safety in Health Care, 14*(6), 443–449.

Gregg, A. C. (1994). Relationship among subjective mental workload, experience, and education of cardiovascular critical care RNs. *Nursing Administration Quarterly, 19*(1), 89–90.

Halbesleben, J. R. (2010). The role of exhaustion and workarounds in predicting occupational injuries: A cross-lagged panel study of health care professionals. *Journal of Occupational Health Psychology, 15*(1), 1–16.

Halbesleben, J. R. et al. (2010). Rework and workarounds in nurse medication administration process: Implications for work processes and patient safety. *Health Care Management Review, 35*(2), 124–133.

Hart, S. G. and Staveland, L. E. (1988). Development of the NASA-TLX (task load index): Results of the experimental and theoretical research. In: P. A. Hancock and N. Meshkati (Eds), *Human mental workload.* Amsterdam: Elsevier, pp. 139–183.

Harvey, C. and Stanton, N. A. (2014). Safety in System-of-Systems: Ten key challenges. *Safety science, 70,* 358–366.

Hawkins, R. et al. (2011). A new approach to creating clear safety arguments. In: C. Dale and T. Anderson (Eds), *Advances in Systems Safety.* London: Springer.

Hawkins, R. et al. (2013). Assurance cases and prescriptive software safety certification: A comparative study. *Safety Science, 59,* 55–71.

Haynes, A. B. et al. (2009). A surgical safety checklist to reduce morbidity and mortality in a global population. *N Engl J Med, 360,* 491–499.

Health and Safety Executive (2001). *Reducing Risk: Protecting People.*

Hendrick, H. and Kleiner, B. (2001). *Macroergonomics: An introduction to work system design.* Santa Monica, CA: Human Factors and Ergonomics Society.

Hjort, P. (2007). *Uheldige hendelser i helsevesenet – en lære-, tenke- og faktabok (Adverse events in healthcare – a learning, thinking and – factbook).* Gyldendal akademisk: Oslo.

Hoffman, R. R. and Militello, L. G. (2008). *Perspectives on cognitive task analysis: Historical origins and modern communities of practice.* Boca Raton, FL: CRC Press.

Hoffman, R. R., Crandall, B. and Shadbolt, N. (1998). Use of the critical decision method to elicit expert knowledge: A case study in the methodology of cognitive task analysis. *Human Factors, 40*(2), 254–276.

Holden, R. J., Rivera-Rodriguez, A. J., Faye, H., Scanlon, M. C. and Karsh, B.T. (2013). Automation and adaptation: Nurses' problem-solving behaviour following the implementation of bar-coded medication administration technology. *Cognition, Technology and Work, 15*(3), 283–296.

Hollnagel, E. (2009a). *The ETTO Principle: Efficiency–Thoroughness Trade-Off.* Farnham, UK: Ashgate.

Hollnagel, E. (2009b). The four cornerstones of resilience engineering. In: C. P. Nemeth, E. Hollnagel and S. W. A. Dekker (Eds), *Resilience engineering perspectives Volume 2. Preparation and restoration.* Farnham, UK: Ashgate, pp. 117–133.

Hollnagel, E. (2011a). Epilogue: RAG – the resilience analysis grid. In: E. Hollnagel et al. (Eds), *Resilience engineering in practice. A guidebook.* Farnham, UK: Ashgate.

Hollnagel, E. (2011b). Looking for patterns in everyday clinical work. In: R. L. Wears, E. Hollnagel and J. Braithwaite (Eds), *Resilient health care, vol. 2: The resilience of everyday clinical work.* Farnham, UK: Ashgate, pp. 145–162.

Hollnagel, E. (2012). *FRAM: the Functional Resonance Analysis Method: Modeling Complex Socio-technical Systems.* Farnham, UK: Ashgate.

Hollnagel, E. (2013a). *FRAM – Funktionel Resonans Analyse Metode – en kursushåndbog.* Denmark: Center for Kvalitet.

Hollnagel, E. (2013b). Making health care resilient: From Safety I to Safety II. In: Hollnagel, E., Braithwaite, J. and Wears, R. L. (Eds), *Resilient health care.* Farnham, UK: Ashgate.

Hollnagel, E. (2014a). *Safety I and Safety II. The past and future of Safety management.* Farnham, UK: Ashgate.

Hollnagel, E. (2014b). Is safety a subject for science? *Safety Science, 67,* 21–24.

Hollnagel, E. (2015a). Disaster management, control and resilience. In: A. Masys (Ed), *Disaster management: Enabling resilience.* Switzerland: Springer International Publishing.

Hollnagel, E. (2015b). Looking for patterns in everyday clinical work, in R. L. Wears, E. Hollnagel and J. Braithwaite (Eds), *Resilient health care volume 2: The resilience of everyday clinical work.* Farnham, UK: Ashgate, pp. 123–138.

Hollnagel, E. (2015c). Why is work-as-imagined different from work-as-done? In: R. L. Wears, E. Hollnagel and J. Braithwaite (Eds), *Resilient health care volume 2: The resilience of everyday clinical work.* Farnham, UK: Ashgate, pp. 249–264.

Hollnagel, E. (2016). Prologue: Why do our expectations of how work should be done never correspond exactly to how work is done. In: J. Braithwaite, R. L. Wears and E. Hollnagel (Eds), *Resilient health care III: Reconciling Work-As-Imagined and Work-As-Done.* Farnham, UK: Ashgate.

Hollnagel, E. (2017). *Safety-II in practice: Developing the resilience potentials.* London and New York: Routledge.

Hollnagel, E., Braithwaite, J. and Wears, R. L. (2013a). Preface: On the need for Resilience in health care. In: E. Hollnagel, J. Braithwaite and R. L. Wears (Eds), *Resilient health care.* Farnham, UK: Ashgate.

Hollnagel, E., Braithwaite, J. and Wears, R. L. (2013b). Epilogue: How to make health care resilient. In: E. Hollnagel, J. Braithwaite and R. L. Wears (Eds), *Resilient health care*. Farnham, UK: Ashgate.

Hollnagel, E., Braithwaite, J. and Wears, R. L. (2013c). *Resilient health care*. Farnham, UK: Ashgate.

Hollnagel, E., Wears, R. L. and Braithwaite, J. (2015). From Safety-I to Safety-II: A White Paper. http://resilienthealthcare.net/onewebmedia/WhitePaperFinal.pdf (accessed 23 December 2015).

Hollnagel, E., Woods, D. D. and Leveson, N. (Eds), (2006). *Resilience engineering: Concepts and precepts*. Farnham, UK: Ashgate.

Holmstrom, J., Ketokivi, M. and Hameri, A. (2009). Bridging practice and theory: A design science approach. *Decision Sciences, 40*(1), 65–87.

Howard, D. (2003). The basics of deployment flowcharting and process mapping. *Management – New Style, 5*.

Hunte, G. S. (2015). A lesson in resilience: the 2011 Stanley Cup Riot. In: R. L. Wears, E. Hollnagel and J. Braithwaite (Eds), *Resilient health care, vol. 2: The resilience of everyday clinical work*. Farnham, UK: Ashgate, pp. 1–10.

Hunte, G. S. and Wears, R. L. (2016). *Power and resilience*. Boca Raton, FL: CRC Press.

Hunter, C. (2008). 'Untangling the web of critical incidents': Ethnography in a paediatric setting. *Anthropology and Medicine, 15*(2), 91–103.

IKAS (2012) Patientidentifikation, standard 1.2.7, I: DDKM, 2. version.

ISMP – Institute for Safe Medication Practices. (2010). CMS 30-minute rule for drug administration needs revision. *Medication Safety Alert!, 15*(18), 1–6.

Jha, A. K. et al. (2010). Patient safety research: An overview of global evidence. *Quality and Safety in Health Care, 19*, 42–47.

Johnson, A. and Lane, P. (2016). Resilience Work-As-Done in everyday clinical work. In: J. Braithwaite, R. L. Wears and E. Hollnagel (Eds), *Resilient health care, vol. 3: Reconciling work-as-imagined with work-as-done*. Farnham, UK: Ashgate.

Johnson, J., Farnan, J. M. et al. (2012). Searching for the missing pieces between the hospital and primary care: mapping the patient process during care transitions. *BMJ Quality & Safety, 21*, i97–i105.

Johnson, K., Geis, G. et al. (2012). Simulation to implement a novel system of care for paediatric critical airway obstruction. *Archives of Otolaryngology – Head & Neck Surgery, 138*(10), 907–911.

Joint Commission on Accreditation of Healthcare Organizations. (1998). Medication error prevention: potassium chloride. *Sentinel Event Alert, 1*.

Jordon, M. et al. (2010). Implications of complex adaptive systems theory for interpreting research about health care organizations. *Journal of Evaluation in Clinical Practice, 16*(1), 228–231.

Jowsey, T. (2015). Watering down ethnography. *BMJ Qual Saf Published Online First*: 30 December 2015.

Kemmis, S. and McTaggart, R. (2003). Participatory action research. In: N. K. Denzin and Y. S. Lincoln (Eds), *Strategies of qualitative inquiry, Vol. 2*. Thousand Oaks, CA: Sage, pp. 336–396.

Kenney, C. (2011). *Transforming health care: Virginia Mason Medical Center's pursuit of the perfect patient experience*. Boca Raton, FL: CRC Press.

Kerr, M. (2011). *Inpatient care for people with diabetes: The economic case for change.* Leicester, UK: NHS Diabetes.

Khan, F. A. and Hoda, M. Q. (2005) Drug related critical incidents. *Anaesthesia, 60* (1), 48–52.

Kitzinger, J. (1996). Focus groups. In: C. Pope and N. Mays (Eds), *Qualitative research in health care.* Malden, MA: Blackwell Publishing.

Klein, G. A., Calderwood, R. and Macgregor, D. (1989). Critical Decision Method for eliciting knowledge. *IEEE Transactions on Systems, Man, and Cybernetics, 19* (3), 462–472.

Kneebone, R. (2003). Simulation in surgical training: Educational issues and practical implications. *Medical Education, 37*(3), 267–277.

Knox, P. (2013). *Who's in charge here? A literature review of approaches to leadership in humanitarian operations.* London: ALNAP/ODI.

Kobayashi, M. etal. (2005). Work coordination, workflow, and workarounds in a medical context. *Conference on Human Factors in Computing Systems: CHI '05,* Portland, OR. New York, NY: ACM.

Koppel, R. et al. (2008). Workarounds to barcode medication administration systems: Their occurrences, causes, and threats to patient safety. *Journal of the American Medical Informatics Association, 15*(4), 408–423.

Kunzle, B. et al. (2010). Leadership in anaesthesia teams: The most effective leadership is shared. *Quality and Safety in Health Care, 19,* 1–6.

Kvale, S. (2002). *Interview,* 1. udgave. Denmark: Reitzel.

Lalley, C. (2014). Workarounds and obstacles: Unexpected source of innovation. *Nursing Administration Quarterly, 38*(1), 69–77.

Langley, A. (1999). Strategies for theorizing from process data. *The Academy of Management Review, 24*(4), 691–710.

Laugaland, K., Aase, K. and Waring, J. (2014). Hospital discharge of the elderly – an observational case study of functions, variability and performance-shaping factors. *BMC Health Services Research, 14,* 365.

Leveson, N. (2011). The use of safety cases in certification and regulation. *Journal of System Safety, 47.*

Lévi-Strauss, C. (1966). *The savage mind.* Chicago, IL: University of Chicago Press.

Lewin, K. (1952). Group decision and social change. In: E. Newcombe and R. Harley (Eds), *Readings in social psychology.* New York: Henry Holt, pp. 459–473.

Lundberg, J. and Johansson, B. (2015). Systemic Resilience Model. *Reliability Engineering and System Safety, 141,* 22–32.

MacIntosh, J. et al. (2012). *The impact of the 22nd February 2011 Earthquake on Christchurch Hospital.* New Zealand Society for Earthquake Engineering (NZSEE) conference proceedings, Christchurch, New Zealand.

Macrae, C. (2013). Reconciling regulation and resilience in health care. In: E. Hollnagel, J. Braithwaite and R. L. Wears (Eds), *Resilient health care.* Farnham, UK: Ashgate.

Maguire, R. (2006). *Safety cases and safety reports.* Farnham, UK: Ashgate.

Malec, J. et al. (2007). The Mayo high performance teamwork scale: Reliability and validity for evaluating key crew resource management skills. *Simulation in Healthcare: The Journal of the Society for Medical Simulation, 2*(1), 4–10.

Malterud, K. (2011). *Kvalitative metoder i medisinsk forskning – en innføring. (Qualitative methods in medical research – an introduction)*. Universitetsforlaget: Oslo.

Manser, T. (2009). Teamwork and patient safety in dynamic domains of healthcare: A review of the literature. *Acta Anaesthesiol Scand, 53*, 143–151.

Marchal, B. et al. (2013). Studying complex interventions: Reflections from the FEMHealth project on evaluating fee exemption policies in West Africa and Morocco. *BMC Health Serv Res, 13*(1), 1–9.

Marshall, C. and Rossman, G. B. (2010). *Designing qualitative research*. Thousand Oaks, CA: Sage.

McAlearney, A. S. et al. (2007). Strategic work-arounds to accommodate new technology: The case of smart pumps in hospital care. *Journal of Patient Safety, 3* (2), 75–81.

McCallin, A. (2003). Interdisciplinary team leadership: A revisionist approach for an old Problem? *Journal of Nursing Management, 11*, 364–370.

McLeod, M., Barber, N. and Franklin, B. D. (2015). Facilitators and barriers to safe medication administration to hospital inpatients: A mixed methods study of nurses' medication administration processes and systems (the MAPS Study). *PloS one, 10*(6), e0128958.

McLuhan, M. (1964). *Understanding media, the extensions of man*. New York, NY: McGraw-Hill.

McNab, D., Bowie, P., Morrison, J. and Ross, A. (2016). Understanding patient safety performance and educational needs using the 'Safety-II' approach for complex systems. *Education for Primary Care, 27*(6), 443–450.

Merton, R. K. (1936). The unanticipated consequences of social action. *American Sociological Review, 1*, 894–904.

Midgley, G. (2003). Science as systemic intervention: Some implications of systems thinking and complexity for the philosophy of science. *System Practice and Action Research, 16*(2), 77–96.

Mielonen, J. (2011). *Making sense of shared leadership. A case study of leadership processes without formal leadership structure in team context*. Finland: Digipaino.

Ministry of Health, Labor and Welfare. (2004). Safety measures to prevent cognitive mix-up in handling high risk medications. *Pharmaceuticals and Medical Devices Safety Information, 202* (in Japanese).

Ministry of Healthcare Services (2008–2009). *A joyous occasion. About a cohesive pregnancy, birth and postnatal care*. Healthcare Ministry: Oslo.

Ministry of Healthcare Services (2010–2011). *National Health Plan 2011–2015. Report to parliament*. Ministry of Healthcare Services: Oslo.

Ministry of Healthcare Services (2012–2013). *Good quality – safe services. Quality and Patient Safety in Healthcare*. Ministry of Healthcare Services: Oslo.

Ministry of Healthcare Services (2014–2015). *Quality and Patient safety*. Ministry of Healthcare Services: Oslo.

Mohr, J. and Arora, V. (2004). Break the Cycle: Rooting out the Workaround. *ACGME Bulletin*, 6–7 November.

Morath, J. and Turnbull, J. (2005). *To do no harm*. San Francisco, CA: Jossey-Bass.

Morgan, D. (1996). Focus Groups. *Annual Review of Sociology, 22*, 129–152.

Naidoo, L. J. et al. (2012). The 2 × 2 model of goal orientation and burnout: The role of approach–avoidance dimensions in predicting burnout. *Journal of Applied Social Psychology, 42,* 2541–2563.

National Patient Safety Agency. (2002). *Patient Safety Alert,* 23 July.

Nawaz, H. et al. (2014). Teaming: An approach to the growing complexities in health care. *The Journal of Bone & Joint Surgery, 96*(21), e184.

Nembhard, I. M. and Edmondson, A. C. (2006). Making it safe: The effects of leader inclusiveness and professional status on psychological safety and improvement efforts in health care teams. *Journal of Organizational Behavior, 27,* 941–966.

Nemeth, C. P. et al. (2006). Creating resilient IT: How the sign-out sheet shows clinicians make healthcare work. *AMIA Annu Symp Proc,* 584–588.

Novak, J. D. and Canas, A. J. (2006). *The theory underlying concept maps and how to construct and use them.* Florida: IHMC Technical Report.

NSW Health (2009). *Respecting patient privacy and dignity in NSW Health.* Sydney: NSW Department of Health.

Nugus, P. et al. (2014). The emergency department 'carousel': An ethnographically-derived model of the dynamics of patient flow. *International emergency nursing, 22* (1), 3–9.

O'Donoghue, T. and Punch, K. (2003). *Qualitative educational research in action: Doing and reflecting.* London: Routledge Falmer.

Orlikowski, W. J. (2002). Knowing in practice: Enacting a collective capability in distributed organizing. *Organization Science, 13*(3), 249–273.

Ormerod, P. (1996). *The death of economics.* New York: St. Martin's Press.

Owen, H (2008). *Open Space Technology: A user's guide* (3rd edn). San Francisco, CA: Berrett-Koehler Publishers.

Paries J. (2006). Complexity, emergence, resilience. In: E. Hollnagel, D. D. Woods and N. Leveson (Eds), *Resilience engineering: Concepts and precepts.* Farnham, UK: Ashgate, pp. 43–53.

Patterson, E. S. et al. (2006). Compliance with intended use of bar code medication administration in acute and long-term care: an observational study. *Human Factors, 48,* 15–22.

Perrow, P. (1999) *Normal accidents: Living with high-risk technologies.* Princeton University Press, Princeton, 1999. Reprint. Originally published: New York: Basic Books, 1984.

Pittet, D. et al. (2009). The World Health Organization Guidelines on Hand Hygiene in Health Care and their consensus recommendations. *Infect Control Hosp Epidemiol, 30,* 611–622.

Pronovost, P. et al. (2006). An intervention to decrease catheter-related bloodstream infections in the ICU. *N Engl J Med, 355,* 2725–2732.

Pronovost, P. et al. (2015). *Transforming patient safety. A sector-wide systems approach.* Report of the WISH Patient Safety Forum 2015. World Innovation Summit for Health.

Pronovost, P. J., Miller, M. and Wachter, R. M. (2006). Tracking progress in patient safety: An elusive target. *Journal of American Medical Association, 296*(6), 696–699.

Rack, L., Dudjak, L. and Wolf, G. (2012). Study of nurse workarounds in a hospital using a code medication administration system. *Journal of Nursing Care Quality, 7* (3), 232–239.

Rankin, A. et al. (2014). Resilience in everyday operations a framework for analyzing adaptations in high-risk work. *Journal of Cognitive Engineering and Decision Making, 8*(1), 78–97.

Rankin, A., Lundberg, J. and Woltjer, R. (2011). Resilience Strategies across industries for managing everyday risks. In: *4th Resilience Engineering International Symposium*, 8–10 June 2011, Sophia Antipolis, France.

Rapport, F. et al. (2017). The struggle of translating science into action: Foundational concepts of implementation science. *Journal of Evaluation in Clinical Practice*, 1–10.

Rasmussen, J. (1997). Risk management in a dynamic society: A modelling problem. *Safety Science, 27*(2–3), 183–213.

Rayo, M. et al. (2007). Assessing medication safety technology in the intensive care unit. *51st Annual Meeting of the Human Factors and Ergonomics Society*, HFES 2007, October 1–5,Baltimore, MD, United States, Human Factors an Ergonomics Society.

Reader, T. W. et al. (2011). Team situation awareness and the anticipation of patient progress during ICU rounds. *BMJ Qual Saf, 20*, 1035–1042.

Reason, J. (1997). *Managing the risks of organizational accidents*. Farnham, UK: Ashgate.

Reason, J. (2000). Safety paradoxes and safety culture. *Injury Control & Safety Promotion, 7*(1), 3–14.

Reason, J. (2004). Beyond the organisational accident: The need for 'error wisdom' on the frontline. *Quality and Safety in Health Care, 13*(2), ii28–2833.

Reason, P. and Bradbury, H. (2001). *Handbook of action research: Participative inquiry and practice*. Centre for action research in professional practice.

Rego, L. and Garau, R. (2008). *Stepping into the void*. Centre for Creative Leadership. Greensboro. North Carolina.

Reynolds, L. R. (2007). An institutional process to improve inpatient glycemic control. *Qual Health Care, 16*(3), 239–249.

Richardson, S. and Ardagh, M. (2013). Innovations and lessons learned from the Canterbury earthquakes. *Disaster prevention and Management, 22*(5), 405–414.

Righi, A. W. and Saurin, T. A. (2015). Complex socio-technical systems: Characterization and management guidelines. *Applied Ergonomics, 50*, 19–30.

Righi, A. W., Saurin, T. A. and Wachs, P. (2015). A systematic literature review of resilience engineering: Research areas and a research agenda proposal. *Reliability Engineering and System Safety, 141*, 142–152.

Ritchie. J. et al. (2013). *Qualitative research practice: A guide for social science students and researchers*. London: Sage.

Robert, G. B. et al. (2011). A longitudinal, multi-level comparative study of quality and safety in European hospitals: The QUASER study protocol. *BMC Health Service Researchs, 11*, 285.

Robson, R. (2015). ECW in Complex Adaptive Systems. In: R. L. Wears, E. Hollnagel and J. Braithwaite (Eds), *Resilient health care volume 2. The resilience of everyday clinical work*. Farnham, UK: Ashgate, pp. 177–188.

Rochlin, G. I. (1999). Safe operation as a social construct. *Ergonomics, 42*(11), 1549–1560.

Rogers, P. J. and Fraser, D. (2003). Appreciating appreciative inquiry. *New Directions for Evaluation, 100*, 75–83.

Roman, S. and Chassin, M. (2001). Windows of opportunity to improve diabetes care when patients with diabetes are hospitalized for other conditions. *Diabetes Care*, *24*(8), 1371–1376.

Ross, A. J. et al. (2014). Inpatient diabetes care: Complexity, resilience and quality of care. *Cogn Technol Work*, *16*(1), 91–102.

Rother, M. and Shook, J. (1998). *Learning to see: Value stream mapping to ADC value and eliminate muda*. Cambridge, MA: The Lean Enterprise Institute.

Saurin, T. A., Rooke, J. and Koskela, L. (2013). A complex systems theory perspective of lean production. *International Journal of Production Research*, *51*, 5824–5838.

Saurin, T. A., Rosso, C. B. and Colligan, L. (2016). Towards a resilient and lean healthcare. In: J. Braithwaite, R. L. Wears and E. Hollnagel (Eds), *Resilient health care Vol 3: Reconciling work-as-imagined and work-as-done*. Burlington, VT: Ashgate.

Schatzki, T. R., Knorr-Cetina, K. and von Savigny, E. (Eds.), (2001). *The practice turn in contemporary theory*. London and New York: Routledge.

Schoville, R. R. (2009). Work-arounds and artifacts during transition to a computer physician order entry: What they are and what they mean. *Journal of Nursing Care Quality*, *24*(4), 316–324.

Schulman, P. R. (1993). The negotiated order of organizational reliability. *Administration & Society*, *25*(3), 353–372.

Seddon, J. (2005). *Freedom from command and control* (2nd edn). Buckingham, UK: Vanguard Ed.

Seddon, M. E. et al. (2014). From ICU to hospital-wide: Extending central line associated bacteraemia (CLAB) prevention. *N Z Med J*, *127*, 60–71.

Senge, P. et al. (1994). *The fifth discipline fieldbook: Strategies and tools for building a learning organization*. New York, NY: Doubleday.

Sewell, W. H., Jr. (1992). A theory of structure: Duality, agency, and transformation. *American Journal of Sociology*, *98*, 1–29.

Siassakos, D. et al. (2013). What makes maternity teams effective and safe? Lessons from a series of research on teamwork, leadership and team training. *Acta Obstetricia et Gynecologica Scandinavica*, *92*, 1239–1243.

Simon, H. A. (1956). Rational choice and the structure of the environment. *Psychological Review*, *63*(2), 129–138.

Sittig, D. F. and Singh, H. (2010). A new sociotechnical model for studying health information technology in complex adaptive healthcare systems. *Quality and Safety in Health Care*, *19*(3), i68–i74.

Smith, D. K. et al. (2009). A study of perioperative hyperglycemia in patients with diabetes having colon, spine, and joint surgery. *J Perianesth Nurs*, *24*(6), 362–369.

Sofaer, S. (1999). Qualitative methods: what are they and why use them? *Health Services Research*, *34*(5/2), 1101–1118.

Spear, S. and Bowen, H. K. (1999). Decoding the DNA of the Toyota production system. *Harvard Business Review*, *77*, 96–108.

Spear, S. J. and Schmidhofer, M. (2005). Ambiguity and workarounds as contributors to medical error. *Annals of Internal Medicine*, *142*(8), 627–630.

Spradley, J. (1980). *Participant observation*. New York, NY: Holt, Rinehart and Winston.

St. Pierre et al. (2011). *Crisis management in acute care settings: Human factors, team psychology, and patient safety in a high stakes environment.* London: Springer.

Staber, U. and Sydow, J. (2002). Organizational adaptive capacity: A structuration perspective. *Journal of Management Inquiry, 11*(4), 408–424.

Starks, H. and Trinidad, S. B. (2007). Choose your method: A comparison of phenomenology, discourse analysis, and grounded theory. *Qualitative Health Research, 17*(10), 1372–1380.

Sujan, M. (2015). An organisation without a memory: A qualitative study of hospital staff perceptions on reporting and organisational learning for patient safety. *Reliability Engineering and System Safety, 144*, 45–52.

Sujan, M. and Furniss, D. (2015). Organisational reporting and learning systems: Innovating inside and outside of the box. *Clinical Risk, 21*, 7–12.

Sujan, M. A. (2012). A novel tool for organisational learning and its impact on safety culture in a hospital dispensary. *Reliability Engineering and System Safety, 101*, 21–34.

Sujan, M. A. et al. (2011). Hassle in the dispensary: Pilot study of a proactive risk monitoring tool for organisational learning based on narratives and staff perceptions. *BMJ Qual Saf, 20*, 549–556.

Sujan, M. A. et al. (2013). Safety cases for medical devices and health IT: Involving healthcare organisations in the assurance of safety. *Health Informatics Journal, 19*, 165–182.

Sujan, M. A. et al. (2014). Clinical handover within the emergency care pathway and the potential risks of clinical handover failure (ECHO): Primary research. *Health Serv Deliv Res, 2*.

Sujan, M. A. et al. (2015a). The development of safety cases for healthcare services: practical experiences, opportunities and challenges. *Reliability Engineering and System Safety, 140*, 200–207.

Sujan, M. A. et al. (2015b). Emergency Care Handover (ECHO study) across care boundaries: The need for joint decision making and consideration of psychosocial history. *Emergency Medicine Journal, 32*, 112–118.

Sujan, M. A. et al. (2015c). Managing competing organizational priorities in clinical handover across organizational boundaries. *Journal of Health Services Research & Policy, 20*, 17–25.

Sujan, M. A. et al. (2016). Should healthcare providers do safety cases? Lessons from a cross-industry review of safety case practices. *Safety Science, 84*, 181–189.

Sujan, M. A. and Felici, M. (2012). Combining failure mode and functional resonance analyses in healthcare settings. *Computer Safety, Reliability, and Security*, 364–375.

Sujan, M. A., Koornneef, F. and Voges, U. (2007). Goal-based safety cases for medical devices: Opportunities and challenges. *Computer Safety, Reliability, and Security*, 14–27.

Sujan, M. A., Rizzo, A. and Pasquini, A. (2002). Contradictions and critical issues during system evolution. *ACM symposium on Applied Computing.* ACM.

Sujan, M., Pozzi, S. and Valbonesi, C. (2016). Reporting and learning: From extraordinary to ordinary. In: J. Braithwaite, R. L. Wears and E. Hollnagel (Eds), *Resilient health care, vol. 3: Reconciling Work-as-Imagined with Work-as-Done.* Farnham, UK: Ashgate.

Sujan, M., Spurgeon, P. and Cooke, M. (2015a). The role of dynamic trade-offs in creating safety – A qualitative study of handover across care boundaries in emergency care. *Reliability Engineering and System Safety*, *141*, 54–62.

Sujan, M., Spurgeon, P. and Cooke, M. (2015b). Translating tensions into safe practices through dynamic trade-offs: The secret second handover. In: R. L. Wears, E. Hollnagel and J. Braithwaite (Eds), *Resilient health care, vol. 2. The resilience of everyday clinical work*. Farnham, UK: Ashgate, pp. 11–22.

Sundhedsstyrelsen. (2013). Vejledning til identifikation af patienter og anden sikring mod forvekslinger i sundhedsvæsenset. *VEJ*, 9808. https://www.retsinformation.dk/Forms/R0710.aspx?id=160895.

Szymczak, J. E. (2016). Infections and interaction rituals in the organisation: Clinician accounts of speaking up or remaining silent in the face of threats to patient safety. *Sociology of Health & Illness*, *38*(2), 325–339.

The Japanese Board of Cardiovascular Surgery (2004). *Urgent warning on handling of KCl concentrate solutions and 10% lidocaine hydrochloride injections* (in Japanese).

The Japanese Circulation Society and six other cardiology-related societies. (2004). *Urgent warning on handling of KCL concentrate solutions and 10% lidocaine hydrochloride injections* (in Japanese).

Thomas, D. R. (2006). A general inductive approach for analyzing qualitative evaluation data. *American Journal of Evaluation*, *27*, 237.

Thomas, E. J., Sexton, J. B. and Helmreich, R. L. (2003). Discrepant attitudes about teamwork among critical care nurses and physicians. *Crit Care Med*, *31*, 956–959.

Thompson, J. B. (Ed.) (1981). *Paul Ricoeur. Hermeneutics and the human sciences*. New York, NY: Cambridge University Press.

Törrönen, J. (2002).Semiotic theory on qualitative interviewing using stimulus texts. *Qualitative Research*, *2*(3), 343–362.

Tucker, A. L. (2009). Workarounds and resiliency on the front lines of health care. *Perspectives on Safety*, 2.

Tucker, A. L. (2013). *Workarounds and Resiliency on the Front Lines of Health Care*, available at https://psnet.ahrq.gov/perspectives/perspective/78 (accessed 4 January 2016).

Tucker, A. L. and Edmondson, A. (2002). Managing routine exceptions: A model of nurse problem solving behavior. *Advances in Health Care Management*, *3*, 87–113.

Tucker, A. L. and Edmondson, A. (2003). Why hospitals don't learn from failures: Organizational and psychological dynamics that inhibit system change. *California Management Review*, *45*(2), 55–72.

Tucker, A. L., Edmondson, A. and Spear, S. (2002). When problem solving prevents organizational learning. *Journal of Organisational Change Management*, *15*(2), 122–136.

Urbach, D. R. et al. (2014). Introduction of surgical safety checklists in Ontario, Canada. *N Engl J Med*, *370*, 1029–1038.

van der Beek, D. and Schraagen, J. M. (2015). ADAPTER: Analysing & Developing Adaptability & performance in Teams to enhance resilience. *Reliability Engineering and System Safety*, *141*, 33–44.

van der Haar, D. and Hosking, D. M. (2004). Evaluating appreciative inquiry: A relational constructionist perspective. *Human Relations, 57*(8), 1017–1036.

Vincent, C. (2010). *Patient safety.* Chichester, UK: Wiley-Blackwell.

Vincent, C. and Amalberti, R. (2015). *Safer healthcare: Strategies for the real world.* New York, NY: Springer.

Vincent, C., Neale, G. and Woloshynowych, M. (2001). Adverse events in British hospitals: Preliminary retrospective record review. *British Medical Journal, 322* (7285), 517–519.

Vogelsmeier, A. A., Halbesleben, J. R. and Scott-Cawiezell, J. R. (2008). Technology implementation and workarounds in the nursing home. *Journal of the American Medical Informatics Association, 15*(1), 114–119.

Wachter, R. (2008). *Understanding patient safety.* New York, NY: McGraw Hill Medical.

Wallace, B. and Ross, A. J. (2006). *Beyond human error: Taxonomies and safety science.* Boca Raton, FL: CRC Press.

Wallace, B., Ross, A. J. and Davies, J. B. (2003). Applied hermeneutics and qualitative safety data: The CIRAS project. *Human Relations, 56*(5), 587–607.

Waring, J., McDonald, R. and Harrison, S. (2006). Safety and complexity: interdepartmental relationships as a threat to patient safety in the operating department. *Journal of Health Organisation and Management, 20*(3), 227–242.

Wears, R. L., Hollnagel, E. and Braithwaite, J. (2015a) (Eds), *Resilient health care, vol. 2: The resilience of everyday clinical work.* Farnham, UK: Ashgate.

Wears, R. L., Hollnagel, E. and Braithwaite, J. (2015b). *Preface to The resilience of everyday clinical work.* In: R. L. Wears, E. Hollnagel and J. Braithwaite (Eds), *Resilient health care, vol. 2: The resilience of everyday clinical work.* Farnham, UK: Ashgate, pp. xxvii–xxix.

Wears, R. L. and Sutcliffe, K. M. (forthcoming). *Still not safe: The rise and fall of patient safety.* Oxford: Oxford University Press.

Weick, K. E. (1987). Organizational culture as a source of high reliability. *California Management Review, 29*(2), 112–127.

Weick, K. E. and Sutcliffe, K. M. (2007). *Managing the unexpected: Resilient performance in an age of uncertainty.* San Francisco, CA: Jossey-Bass.

Weinger, M. B., Reddy, S. B. and Slagle, J. M. (2004). Multiple measures of anesthesia workload during teaching and non-teaching cases. *Anesthesia & Analgesia, 98*(5), 1419–1425.

Welp, A., Meier, L. L. and Manser, T. (2016). The interplay between teamwork, clinicians' emotional exhaustion, and clinician-rated patient safety: A longitudinal study. *Crit Care, 20,* 110.

West, M. et al. (2014). *Developing collective leadership for healthcare.* London: Kings Fund.

Westrum, R. (2006). A typology of resilience situations. In: E. Hollnagel, D. D. Woods, and N. Leveson (Eds), *Resilience engineering – concepts and precepts.* Farnham, UK: Ashgate, pp. 55–66.

Wiig, S. et al. (2013). Investigating the use of patient involvement and patient experience in quality improvement in Norway: Rhetoric or reality? *BMC Health Services Research, 13,* 206.

Wolfinger, N. H. (2002). On writing fieldnotes: Collection strategies and background expectancies. *Qualitative Research*, *2*(1), 85–93.

Woods, D. D. (2006). Essential characteristics of resilience. In: E. Hollnagel, D. D. Woods and N. Leveson (Eds), *Resilience engineering: concepts and precepts*. Farnham, UK: Ashgate, pp. 21–34.

Woods, D. D. and Cook, R. I. (2006). Incidents: Markers of resilience or brittleness? In: E. Hollnagel, D. D. Woods and N. Leveson (Eds), *Resilience engineering: Concepts and precepts*. Farnham, UK: Ashgate, pp. 73–75.

Woodward, A. M. (2015). *Tapestry of tears: An autoethnography of leadership, personal transformation, and music therapy in humanitarian aid in Bosnia Herzegovina*. Doctoral dissertation from Antioch University. Retrieved from http://aura.antioch.edu/etds/192.

World Health Organization (WHO). (2004). *World alliance for patient safety: Forward programme*. World Health Organization (WHO): Geneva.

Wright, M. C., Taekman, J. M. and Endsley, M. R. (2004). Objective measures of situation awareness in a simulated medical environment. *Quality & Safety in Health Care*, *13*, 65–71.

Yin, R. K. (2009). *Case study research. Design and methods*. Thousand Oaks, CA: Sage.

Zhuravsky, L. (2015). Crisis leadership in an acute clinical setting: Christchurch Hospital, New Zealand ICU experience following the February 2011 earthquake. *Prehospital and Disaster Medicine*, *30*(2).

Zink, B. J. (2005). *Anyone, anything, anytime: A history of emergency medicine*. Philadelphia, PA: Mosby Elsevier.

Index

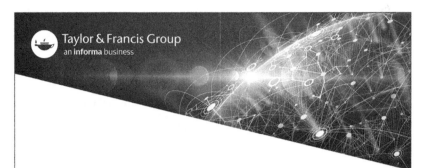

Printed in the United States
by Baker & Taylor Publisher Services